Improving the Energy Efficiency of Mobile Networks

This book provides an analysis of the problem of high energy consumption of mobile networks and describes methods for improving the energy efficiency of mobile networks. It explores new technologies and strategies that will enable mobile networks to operate with less energy and become more sustainable while not compromising service quality.

Improving the Energy Efficiency of Mobile Networks: Strategies for 5G and 6G Technologies is designed to help readers understand how to optimize energy consumption in mobile networks. It explores technical solutions for reducing the energy consumption with a focus on innovative solutions such as network energy consumption optimization using artificial intelligence (AI) in the management of network resources, energy harvesting (EH), software-defined networking, renewable energy sources, and implementation of advanced techniques and protocols for improving the energy efficiency of mobile networks. The book analyzes how the integration of different advanced technologies, network function virtualization, and non-orthogonal multiple access techniques, along with the implementation of AI, renewable energy sources, and EH, can contribute to the optimizing energy use of mobile networks, without compromising network performance. It offers an overview of concrete solutions, with an insight into future directions of development, which makes it particularly relevant at a time when industry is increasingly moving toward sustainability and the implementation of green technologies.

This book is ideal for researchers, industry professionals, and scholars who want to gain a deeper understanding of the excessive mobile network energy consumption and contribute to the development of energy-efficient and sustainable mobile networks.

Improving the Energy Efficiency of Mobile Networks

Strategies for 5G and 6G Technologies

Josip Lorincz
Faculty of electrical engineering, mechanical engineering and naval architecture (FESB), University of Split, Split, Croatia

and

Zvonimir Klarin
Šibenik University of applied science, Šibenik, Croatia

CRC Press
Taylor & Francis Group
Boca Raton London New York

CRC Press is an imprint of the Taylor & Francis Group, an **informa** business

Designed cover image: Shutterstock ©

First edition published 2025
by CRC Press
2385 NW Executive Center Drive, Suite 320, Boca Raton FL 33431

and by CRC Press
4 Park Square, Milton Park, Abingdon, Oxon, OX14 4RN

CRC Press is an imprint of Taylor & Francis Group, LLC

ISBN: 978-1-041-01660-1 (hbk)
ISBN: 978-1-041-01661-8 (pbk)
ISBN: 978-1-003-61586-6 (ebk)

DOI: 10.1201/9781003615866

Typeset in Palatino
by Deanta Global Publishing Services, Chennai, India

Contents

About the Authors

Josip Lorincz (ORCID ID: 0000-0001-7691-582X, WoS ID: D-8869-2017) received his BSc (MS equivalent) and PhD degree in telecommunications engineering and computer science from the Faculty of electrical engineering, mechanical engineering and naval architecture (FESB), University of Split, Croatia, in 2002 and 2010, respectively. In 2003 he joined the Department of Electronics and Computing at the FESB, University of Split, Croatia, where he currently works as a full professor. He was also a visiting researcher at the Politecnico di Milano University, Milan, Italy. As a project leader or researcher, he participated in more than twenty scientific and professional projects funded by the public or private sector. He is the founder and co-chair of the Symposium on Green Networking and Computing, organized in the frame of the International Conference on Software, Telecommunications and Computer Networks (SoftCOM). His current research interests include energy-efficient wireless and wired networks, optimization in telecommunications, advanced design, the management and analyses of computer heterogeneous networks, and performance evolution of routing protocols. He is a senior IEEE member, a senior ACM member, associate member of the Croatian Academy of Engineering, and the first president of the Croatian ACM chapter.

Zvonimir Klarin (ORCID ID: 0000-0003-4855-4128, WoS ID: AAA-4553-2022) currently serves as a lecturer at the Šibenik University of Applied Sciences, Croatia, where he teaches various courses related to computer science. He received his MS degree in Computer Science from the Faculty of Electrical Engineering, Mechanical Engineering, and Naval Architecture (FESB), University of Split, Croatia, where he is currently pursuing a PhD, specializing in energy efficiency in wireless communication networks. His research focuses on optimizing energy consumption in both wireless and wired networks, telecommunications, improving network efficiency, and the management and analysis of heterogeneous computer networks. He is a member of IEEE and participates in the organization of international scientific and professional conferences.

Preface

The rapid evolution of mobile networks has placed improvement of mobile network energy efficiency at the forefront of research and technological development. With the increasing demand for higher data rates, lower latency, and seamless connectivity, mobile networks have expanded significantly, leading to a rise in energy consumption and environmental concerns. Today, mobile communication networks account for a substantial share of global electricity consumption, with user devices, base stations, and other network elements contributing to a growing carbon footprint. As the world moves toward sustainable digital transformation, optimizing energy use in mobile networks has become a global imperative. Governments, regulatory bodies, and industry stakeholders are emphasizing the need for green and energy-efficient mobile network implementations.

This book provides a comprehensive overview of strategies, technologies, and innovations aimed at improving the energy efficiency of mobile communication networks, particularly in the context of 5G and future 6G technologies. By exploring both foundational principles and emerging trends, this book serves as a guiding resource for addressing the energy sustainability challenges of mobile networks, while ensuring that future mobile networks remain high-performing and resilient.

Intended for scientists, scholars, practitioners, and professionals, this book serves as a valuable resource for understanding the challenges and advancements in energy-efficient mobile network design and implementation. It is structured to provide both theoretical foundations and practical insights, encouraging further research and innovation in this critical area.

The first chapter of the book, "Introduction to the Problem of Excessive Mobile Network Energy Consumption," outlines the key motivations for improving mobile network energy efficiency, addressing environmental sustainability, operational costs, regulatory requirements, and technological advancements.

The second chapter of the book, "Evolution of Mobile Networks," explores the transition from the first-generation (1G) to the future sixth-generation (6G) networks, emphasizing the role of heterogeneous networks, millimetre-wave technologies, and artificial intelligence in shaping next-generation networks.

Chapter 3 of the book, entitled "Energy Efficiency of Mobile Networks," defines the metrics, factors, and standardization efforts related to energy efficiency in mobile communication networks. It also examines the impact of network interference, latency, spectral efficiency, and traffic variations on energy consumption.

Chapter 4 of the book, entitled "Optimization of Mobile Network Energy Consumption," presents various techniques for reducing mobile network energy consumption, including base station sleep mode strategies, intelligent base station transmit power control, massive multiple-input multiple-output technologies, cognitive radio techniques, and cloud-based radio access network architectures. Additionally, Chapter 4 discusses energy harvesting and renewable energy integration in the power supply of mobile networks and the impact of network slicing on the energy efficiency of mobile networks. The chapter ends with analyses of the impact of satellite-based mobile network constellations and reconfigurable intelligent surfaces on energy consumption improvements of mobile networks.

The final book chapter, "Future Challenges and Opportunities," explores emerging trends in mobile network energy management, such as AI-driven network optimization and the sustainability of 6G networks, emphasizing the importance of interdisciplinary collaboration and innovation in shaping the future of sustainable wireless communication. The final chapter provides a forward-looking perspective on the next generation of mobile networks, particularly in the transition toward 6G technologies and the global push toward carbon-neutral mobile networks.

This book is designed as an overview resource, aiming to illuminate new research directions in the field of energy-efficient mobile networks. By bridging theoretical advancements with practical implementations, it provides a foundation for further innovation, ensuring that future networks are not only high-performing but also sustainable.

1

Introduction to the Problem of Excessive Mobile Network Energy Consumption

Regardless of a country's economic status or an industry sector's level of development, enhancing mobile network energy efficiency (EE) contributes to a global reduction in energy consumption (EC) and greenhouse gas (GHG) emissions. Achieving these reductions requires the effective implementation of information and communication technologies, which serve as a fundamental tool for driving EE improvements across all countries or industry sectors worldwide. The rapid proliferation of connected devices in the modern digital era, especially smartphones and Internet of Things (IoT) devices, has resulted in significant changes in the architecture and performance requirements of wireless communication networks. As a consequence, the operators of wireless communication networks now face the challenges of managing the problem of growing EC of mobile cellular networks.

One of the main reasons for this growth of network EC is due to the increase in the number of connected user devices, which also imposes numerous operational challenges for wireless communications networks. Smartphones as an example of typical user devices transfer significant amounts of data through mobile application updates, real-time video streaming, social media services, and other various user activities on the Internet. In addition, IoT devices are constantly collecting, processing, and transmitting data, thus increasing the need for the capacity of mobile networks. The Ericsson Mobility Report [1] predicts that by 2030 the number of IoT devices will reach 42.9 billion, while the number of mobile phone subscribers will exceed 9.5 billion. Furthermore, the global average monthly data transfer per smartphone is expected to reach 21.58 GB in 2025 and is projected to increase to 39.52 GB by 2030, indicating a significant increase in data traffic and the need for extremely large data capacities.

Addtionally, the Ericsson Mobility Report [1] predicts that global mobile data traffic will reach 303 EB per month by 2030. Taking into account the traffic generated by user devices via fixed wireless access (FWA) technology, the predictions estimate that total mobile network traffic will reach 473 EB per month by the end of 2030 (Figure 1.1). The growth in mobile network traffic is mainly attributed to an increase in the amount of video content, which is estimated to account for 74% of total mobile data traffic at the end of 2024.

With such expected exponential growth of connected devices and data volumes by 2030, the practical realization of more energy-efficient mobile

DOI: 10.1201/9781003615866-1

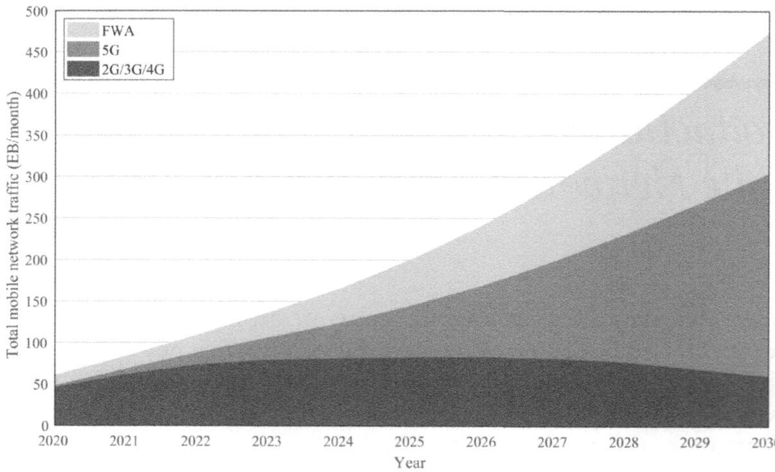

FIGURE 1.1
The projected growth of total global mobile network traffic from 2020 to 2030 [1]

networks becomes challenging. This challenge is primarily related to the need for deployment of additional base stations (BSs), the improvement of existing network EE, and the introduction of new types of services and applications which are mostly more throughput intensive than the previous ones. The main solution for satisfying this challenge is seen in the ongoing deployment of fifth-generation (5G) and future deployment of sixth-generation (6G) mobile networks. Although 5G mobile network technology in comparison with previous generations of mobile networks can provide improved EE per unit of transferred data, the total mobile network EC may increase due to the increase in the number of installed BSs. Namely, each newly installed BS requires additional energy for operation, maintenance, and, in specific implementations, cooling. This challenge thus highlights the importance of matching network EE with the growing capacity requirements of modern mobile networks.

In addition, the nonnegligible GHG emissions of mobile network equipment have also become a concern in the telecommunications sector, which is also driven by the increase in the number of connected devices and the continuous growth of network traffic volumes. As a result, improving mobile network EE and the consequent reduction of GHG emissions have become a priority for mobile network operators. In light of this, the 5G technology can offer significant potential for network EE improvements that can help mobile operators achieve sustainability goals and more effectively reduce the negative impact of the contribution of telecommunication network equipment to global GHG emissions. Thus, the implementation of 5G networks sets new benchmarks for the improvement of network EE through the optimization of network resources and the introduction of specific technologies that can contribute to significant improvements in the optimization of mobile network EC.

Due to the abovementioned reasons, EE has been defined as a critical performance metric for 5G networks and different energy-efficient technologies have been integrated into all recent standards related to the development of 5G technology [2–4]. The mobile network EE is formalized as a key performance indicator, thus imposing new standards in the future development of mobile networks. Accordingly, the behavior of mobile networks in terms of EE requires careful analysis during the planning, deployment, operation, and even disposal phases of the mobile network [2].

The integration of energy-efficient technologies into the standardization of 5G networks offers numerous benefits. One of the main benefits is related to the reduction of operational expenditure (OPEX) costs for telecommunications service providers, which are directly reduced by decreasing network equipment EC and can, in turn, lead to lower service prices for end users. Furthermore, reducing GHG emissions through the improvement of mobile network EE contributes to the mitigation of negative environmental impacts of GHG emissions.

In addition to the ongoing implementation of 5G networks, the upcoming 6G mobile network technology will further enhance network performance, and it is expected to additionally contribute to the improvement of mobile network EE [5, 6]. Compared to 5G networks, the 6G networks are expected to bring faster data rates, reduced latency, and increased capacity [7]. Besides these benefits, the 6G networks are expected to have higher network EE compared to 5G networks, which will be achieved through the use of artificial intelligence (AI)-based solutions dedicated to the optimization of network EE that can thus significantly reduce the negative environmental impact of mobile networks. However, these goals will be challenged with enabling new applications and use cases that are expected to be characteristic for 6G networks, such as the implementation of services related to autonomous driving and robotics, unmanned aerial vehicles, virtual reality, augmented reality, and the Internet of Everything in everyday life. This will consequently result in a continuation of the increase in demand for data capacity, as all of these applications require the transmission of large amounts of data due to their complexity and resource-intensive nature. Therefore, issues related to the improvement of mobile network EE will continue even with the proliferation of 6G networks, and special attention will need to be paid to optimization of network EE, in order to optimize network EC and keep GHG emissions at acceptable levels [8].

Considering the presented advantages that new generations of mobile networks bring and issues related to the problem of excessive EC, it is clear that improving EE in 5G and future 6G networks has considerable potential to positively impact both society and the environment. Hence, improving the EE of mobile networks will be crucial for addressing excessive energy consumption and GHG emissions of mobile networks. However, with the support of 5G and 6G technologies, mobile network operators will need to meet the growing demands for high data transfer rates and capacity and, at the same time, ensure the sustainable development and deployment of current and upcoming generations of mobile networks.

1.1 Motivation for Improving Mobile Network Energy Efficiency

Research in the field of improving EE of wireless communication networks started intensively during the 2010s. It was motivated by increased environmental awareness related to the evident climate change, the need for sustainable development of society, and the necessity for reducing OPEX of mobile network operators. This was followed by national, regional, or global regulatory demands that started to include improvement of mobile network EE in their recommendations and standards. Hence, these motivations foster technological advancements that will result in the development of new hardware solutions, algorithms, techniques, and strategies that are dedicated to improving mobile network EE, while simultaneously meeting the demands for increasing network capacity, ensuring quality of service (QoS) and performance requirements of wireless communication networks.

1.1.1 Environmental Awareness and Sustainability

As awareness of climate change and the need for sustainable development rises, there is increasing pressure on mobile network operators (MNOs) and network equipment manufacturers to reduce GHG emissions and energy consumption of mobile networks. Improving the EE of mobile communication networks is a key step toward achieving these goals. Climate change represents one of the most important global challenges of our time, and an increasing number of scientific studies and reports emphasize its seriousness and the need for urgent intervention [9]. Growing awareness of climate change has increased demands on industries, especially the telecommunications sector, to take measures to reduce their impact on the environment [10]. This transition is forcing the telecommunications sector to adopt sustainable practices and technologies that reduce these negative environmental impacts.

1.1.2 Operating Expenses

Energy accounts for a substantial portion of the OPEX of mobile network operators. According to a GSMA report [11], energy accounts for an average of 20%–40% of total OPEX in the telecommunications sector. This is why the telecommunications industry must pay close attention to reducing its overall EC in order to increase competitiveness and achieve global economic sustainability. The continuous power supply of BSs, cooling systems, data centers, and other networking equipment that enable mobile networks to operate generates high energy costs. Furthermore, the continuous demands for ensuring faster data rates and the emergence of more throughput-demanding services additionally contribute to the increase in the energy requirements of mobile networks, leading to higher OPEX of MNOs. Also, mobile network designs based on the installation of a vast amount of new equipment when deploying every new generation of mobile networks, whose operation also relies on constant, uninterrupted power supply sources, significantly contribute to the increase in capital expenditures (CAPEX) of MNOs. This presents an important challenge for the strategic financial planning of network service providers. In the future, this problem will be further emphasized by the deployment of ultra-dense 5G networks and the expected implementation of 6G networks, which will require an increase in the number of BSs and significantly faster data processing capabilities of mobile network equipment [12, 13]. Consequently, the increasing complexity and amount of mobile network infrastructure will contribute to the further increase of network EC and MNOs' OPEX, forcing MNOs to seek more sustainable and energy-efficient solutions for the realization of radio, backhaul, and core parts of mobile networks.

Given the competitive market environment, MNOs are facing pressure to reduce OPEX in order to remain profitable and competitive in the telecom market. Thus, increasing mobile network EE has become one of the key strategies for reducing OPEX, enabling MNOs to optimize their operations and increase profitability. Ultimately, the reduction of OPEX through the optimization of network EE strengthens the economic resilience and competitiveness of mobile operators on the global market.

1.1.3 Regulatory Requirements

More and more countries are introducing stricter regulations related to energy consumption and GHG emissions. Mobile operators must comply with these requirements to avoid sanctions and additional costs associated with non-compliance. For example, the European Union has introduced a series of regulations to reduce greenhouse gas emissions, including a target of reducing emissions by 40% by 2030 compared to 1990 levels [14]. This decision is part of a broader strategy that also includes increasing the share of renewable energy and improving energy efficiency in order to achieve climate neutrality by 2050.

In the United States, many states are also adopting targets to reduce greenhouse gas emissions and implementing market-based policies. According to a report by the National Conference of State Legislatures, at least 16 states have adopted legislative goals to reduce GHG emissions [15]. These goals often include emissions monitoring and reporting systems, as well as a cap-and-trade system to regulate emissions in the energy and transportation sectors. These efforts are aimed at reducing GHG emissions and encouraging the use of renewable energy sources, thus supporting the broader goals of sustainability and environmental protection.

Additionally, improving mobile network energy efficiency aligns with several of the United Nations Sustainable Development Goals [16], which include a contribution to:

- Goal 7: affordable and clean energy – by reducing the energy consumption per unit of data transmitted and thus lowering GHG emissions and promoting sustainable energy use,
- Goal 9: industry, innovation, and infrastructure – through integrating advanced, energy-saving technologies like renewable-powered base stations and efficient network design,
- Goal 11: sustainable cities and communities goal – by reducing the carbon footprint of urban telecommunications infrastructure, supporting cleaner and more sustainable cities,
- Goal 12: responsible consumption and production – by conserving network resources and encouraging telecom operators to adopt sustainable practices, from manufacturing to operations and disposal, and
- Goal 13: climate action – by reducing GHG emissions which enables mitigating climate change effects and aligning with national and international climate goals.

The telecommunications sector must adapt to such a regulatory framework, through reducing the EC and GHG emissions of mobile networks. However, the introduction of green policies and energy-efficient technologies in mobile networks is becoming imperative not only due to regulatory requirements but also due to MNOs' economic profitability and environmental sustainability. Hence, the goal dedicated to optimizing EE of wireless mobile networks is not only a response to regulatory requirements but also a key step toward long-term environmental preservation and reduction of OPEX.

1.1.4 Technological Advancement

Technological progress plays a key role in increasing the EE of wireless communication networks. The development of new technologies in the frame of 5G and the upcoming 6G networks can bring significant changes in the way networks are designed and their EC is managed [17, 18]. It is expected that these technologies will enable the implementation of advanced solutions that optimize mobile network EC and reduce OPEX. Technological progress enables the possibility of using sophisticated systems that can ensure more precise management of mobile network EC. This will be particularly fostered through the development of new AI-based solutions for dynamic adjustment of network resources in real time, which can result in more energy-efficient operation of mobile networks. Considering the growing demands for higher network capacity and improved performance, technological progress enables EE to be achieved without compromising the QoS. Such technological innovations are crucial for the sustainable development of wireless communication networks and the reduction of their ecological footprint.

References

1. F. Jejdling, *Ericsson Mobility Report*, Stockholm, Sweden: Telefonaktiebolaget LM Ericsson, Nov. 2024.
2. European Telecommunications Standards Institute ETSI, *ES 203 228 (V1.3.1) Environmental Engineering (EE); Assessment of mobile network energy efficiency*, Sophia Antipolis, France: ETSI, Oct. 2020.
3. 3GPP, "TR 32.972 (V16.1.0) (2019-09) Study on system and functional aspects of energy efficiency in 5G networks (Release 16)," 2019. Accessed: Aug. 24, 2024. [Online]. Available: https://portal.3gpp.org/desktopmodules/Specifications/SpecificationDetails.aspx?specificationId=3279
4. ITU-T, "L.1331 (01/2022) Assessment of mobile network energy efficiency," 2022. Accessed: Aug. 24, 2024. [Online]. Available: https://www.itu.int/rec/T-REC-L.1331-202201-I/en
5. R. Kumar, S. K. Gupta, H. C. Wang, C. S. Kumari, and S. S. V. P. Korlam, "From efficiency to sustainability: Exploring the potential of 6G for a greener future," 2023. doi: 10.3390/su152316387.
6. Fraunhofer IIS, "6G energy efficiency and sustainability," 2023. Accessed: Aug. 17, 2024. [Online]. Available: https://www.iis.fraunhofer.de/content/dam/iis/en/doc/lv/Whitepaper6GSustainability.pdf
7. J. Hu, Q. Wang, and K. Yang, "Energy self-sustainability in full-spectrum 6G," *IEEE Wirel Commun*, vol. 28, no. 1, 2021. doi: 10.1109/MWC.001.2000156.
8. N. Hu, Z. Tian, X. Du, and M. Guizani, "An energy-efficient in-network computing paradigm for 6G," *IEEE Transactions on Green Communications and Networking*, vol. 5, no. 4, 2021. doi: 10.1109/TGCN.2021.3099804.
9. W. Wu, Z. Xu, S. Zenchenko, and W. Strielkowski, "Does sustainable digital economy have implications for climate change? Novel evidence in the context of sustainable energy technologies," *Sustainable Energy Technologies and Assessments*, vol. 60, 2023. doi: 10.1016/j.seta.2023.103524.
10. Suslina, K. Savin, and I. Suslina, "Greening telecom: Harnessing the power of artificial intelligence for sustainable communications," in *Studies in Computational Intelligence*, 2024. doi: 10.1007/978-3-031-50381-8_94.
11. E. Kolta and T. Hatt, "A blueprint for green networks," GSMA Intelligence, Oct. 2022.

12. M. W. Akhtar, S. A. Hassan, R. Ghaffar, H. Jung, S. Garg, and M. S. Hossain, "The shift to 6G communications: Vision and requirements," *Human-centric Computing and Information Sciences*, vol. 10, no. 1, 2020. doi: 10.1186/s13673-020-00258-2.
13. R. S. Yarrabothu and G. R. Murthy, "Ultra-dense deployments in next-generation networks and metaverse," in *Towards Wireless Heterogeneity in 6G Networks*, 2024. doi: 10.1201/9781003369028-3.
14. UNFCCC, "EU agrees 40% greenhouse gas cut by 2030." Accessed: Jul. 18, 2024. [Online]. Available: https://unfccc.int/news/eu-agrees-40-greenhouse-gas-cut-by-2030
15. L. Shields, "Greenhouse gas emissions reduction targets and market-based policies." Accessed: Jul. 20, 2024. [Online]. Available: https://www.ncsl.org/energy/greenhouse-gas-emissions-reduction-targets-and-market-based-policies
16. United Nations, "The sustainable development goals report 2023 special edition," United Nations, New York, 2023.
17. S. Maiti and S. Juneja, "Energy efficiency techniques in 5G/6G networks: Green communication solutions," in *Lecture Notes in Networks and Systems*, 2024. doi: 10.1007/978-981-99-9524-0_31.
18. T. Huang, W. Yang, J. Wu, J. Ma, X. Zhang, and D. Zhang, "A survey on green 6G network: Architecture and technologies," *IEEE Access*, vol. 7, 2019. doi: 10.1109/ACCESS.2019.2957648.

2

Evolution of Mobile Networks

The development of mobile networks from the first generation (1G) to the fifth generation (5G) represents continuous technological progress that meets the increasing needs for faster and larger data transmissions, higher network capacities, lower data transmission latencies, and advanced network architectures (Table 2.1). Each generation brings significant improvements in data transfer speed, exploitation of frequency channel widths, number of supported services, and technical standardization [1]. While early generations of mobile networks focused on voice and text message transmission, newer generations introduced advanced capabilities such as video streaming, virtual reality (VR), augmented reality (AR), and tactile communications. Over time, the complexity of network architectures has increased, transitioning from homogeneous and simple single information stream systems to heterogeneous systems with data transmission based on multiple-input multiple-output (MIMO) and massive MIMO (mMIMO) technologies. This technological transition of mobile networks brings into focus issues related to network energy efficiency (EE), security, and privacy in upcoming next-generation mobile network technologies. While 5G is currently in the stage of practical implementation on a global scale, the scientific community and industry are already preparing for the evolution toward 6G networks, which will bring even greater changes in how a vast number of user devices (UDs) will be connected in mobile networks.

2.1 5G Mobile Network

The 5G mobile network was developed to significantly improve the performance characteristics of the previous 4G mobile networks in terms of faster data transfer rates, higher capacity, lower latency, and support for new use cases and applications.

A typical 5G mobile network consists of three key components: user device (UD), the core network (CN), and the radio access network (RAN) (Figure 2.1). User equipment includes typical devices such as mobile phones, tablets, fixed wireless access (FWA) devices, and Internet of Things (IoT) devices that end-users use to access the network. In mobile networks, the CN manages network signaling, traffic routing, and service provisioning. In 4G mobile

DOI: 10.1201/9781003615866-2

TABLE 2.1

Comparison of Mobile Network Technologies through Their Evolution in Time [1]

Features	1G	2G	3G	4G	5G	6G (assumed)
Period	1980–1990	1990–2000	2000–2010	2010–2020	2020–2030	2030–2040
Maximum data rate	2.4 Kbit/s	144 Kbit/s	2 Mbit/s	1 Gbit/s	35.46 Gbit/s	100 Gbit/s
Maximum frequency	894 MHz	1,900 MHz	2,100 MHz	6 GHz	90 GHz	10 THz
Service level	Voice	Text	Picture	Video	3D VR/AR	Tactile
Standards	MTS, AMPS, IMTS, PTT	GSM, IS-95, CDMA, EDGE	UMTS, WCDMA, IMT2000, CDMA2000, TD-SCDMA	WiMAX, LTE, LTE-A	5G NR, 5G Advanced	–
Multiplexing	FDMA	FDMA, TDMA	CDMA	OFDMA	OFDMA	NOMA, smart OFDMA plus IM
Architecture	Homogeneous SISO	Homogeneous SISO	Homogeneous SISO	Partially heterogeneous, MIMO	Heterogeneous MIMO	Highly heterogeneous, Intelligent surface, mMIMO
Core network	PSTN	PSTN	Packet-switched network	Internet	Internet, Internet of Things	Internet of Everything
Highlight	Mobility	Digitization	Internet	Real-time streaming	Extremely high data rate	Security, secrecy, privacy

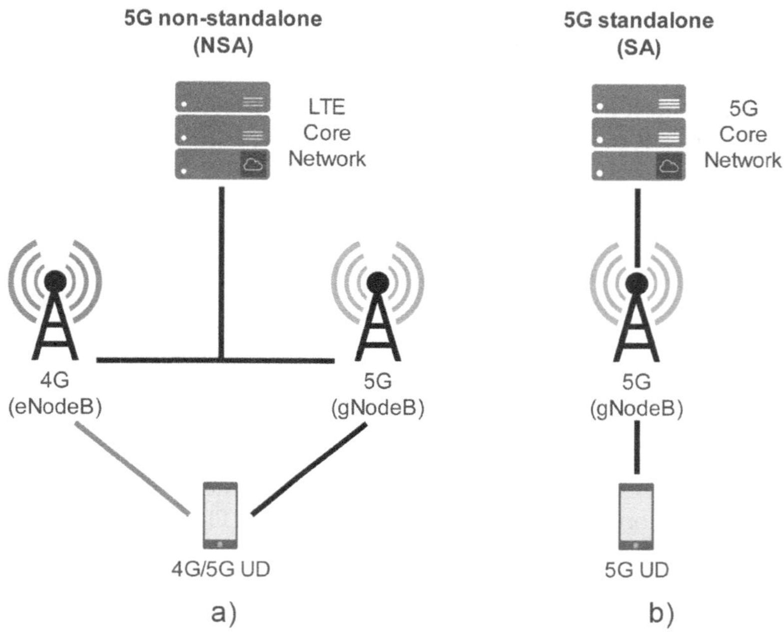

FIGURE 2.1
A comparison of two configurations of 5G mobile networks known as (a) 5G non-standalone (NSA) and (b) 5G standalone (SA) [2]

networks, also known as long-term evolution (LTE) networks, this function is performed by the evolved packet core (EPC) network component (Figure 2.1), a part of the 4G network that enables data transmission, mobility management, and user security [3]. With the advent of 5G mobile networks, the 5G core (5GC) component (Figure 2.1) introduces advanced network functionalities that support ultra-low latency, massive connectivity, and flexible network segmentation using network slicing techniques [4]. These features allow the 5G network to adapt to various types of applications and services, from low-latency autonomous vehicle communications to communications supporting a large number of IoT devices, thus significantly expanding the range of possible applications compared to previous mobile network generations.

The RAN connects UDs to the CN through base stations (BSs), which serve as the interface for providing spatially dispersed users access to various services and applications (Figure 2.1). In 4G networks, BSs are known as evolved Node B (eNodeB) and are responsible for resource management and UD mobility control. In 5G networks, the BSs are called next-generation Node B (gNodeB) and bring additional advanced capabilities compared to eNodeB, including support for higher data transfer rates, lower data transfer delays, and the simultaneous connection with BSs of a much larger number of UDs. These capabilities represent the basis for the practical realization of new services such as massive IoT and autonomous vehicles in 6G networks.

A key role in this evolution is related to the development of 5G new radio (NR), a new communication technology that provides significant improvements in capacity, data transfer speed, and reduction in data transmission latency compared to previous 4G networks [5]. Designed specifically for 5G networks, 5G NR provides a flexible and scalable platform that can support a broad range of applications, from mobile broadband communication to massive device connectivity within the IoT paradigm.

The introduction of 5G technology marked a significant milestone in the evolution of mobile networks, not only because of the significant increase in the speed of data transmission but also because of the possibilities that enable the development of many new network services. The initial deployment of 5G technology was carried out in the form of the so-called non-standalone architecture (NSA) (Figure 2.1a), where the 5G network relies on the existing 4G CN for network signaling and management [6]. This approach allowed for a faster rollout of 5G services and broader coverage during the early stages of implementation. Over time, a complete transition to a 5G standalone (SA) architecture is expected (Figure 2.1b), which uses the new 5G core network and enables complete independence from the 4G infrastructure [4]. The introduction of 5G SA network architecture brings advanced functionalities such as ultra-low latency, greater network flexibility, and support for advanced industrial applications, further enhancing the capabilities of mobile networks.

2.1.1 Increase of Data Traffic and Its Implications

The evolution of 5G network architecture enables the transmission of a significantly larger amount of data traffic than the previous generation (4G) of mobile networks, which is important for satisfying current trends characterized by continuous annual data traffic growth. These trends indicate that by the end of 2024, 5G was projected to account for 34% of total mobile network data traffic, and this share is expected to increase to 80% by 2030 [7]. This increase in 5G data traffic reflects not only the migration of users from previous generations of mobile networks to 5G networks but also the growing adoption of new services and applications that are enabled by the advanced capabilities of 5G networks. While some of this growth is due to the migration of users to newer and more advanced mobile network generations, demand for technologies such as extended reality (XR) is expected to drive a further data traffic increase in mobile networks. Thus, this trend highlights the capabilities of 5G networks and especially future 6G networks to support new use cases and services that were infeasible for practical realization in previous mobile network generations.

Also, 5G networks are projected to become the dominant mobile access technology in terms of the overall number of subscriptions by 2027. It is expected that nearly 6.3 billion 5G subscriptions will be in 5G networks by 2030 (Figure 2.2), which will account for 67% of mobile subscriptions in all mobile network generations. The further development of 5G networks,

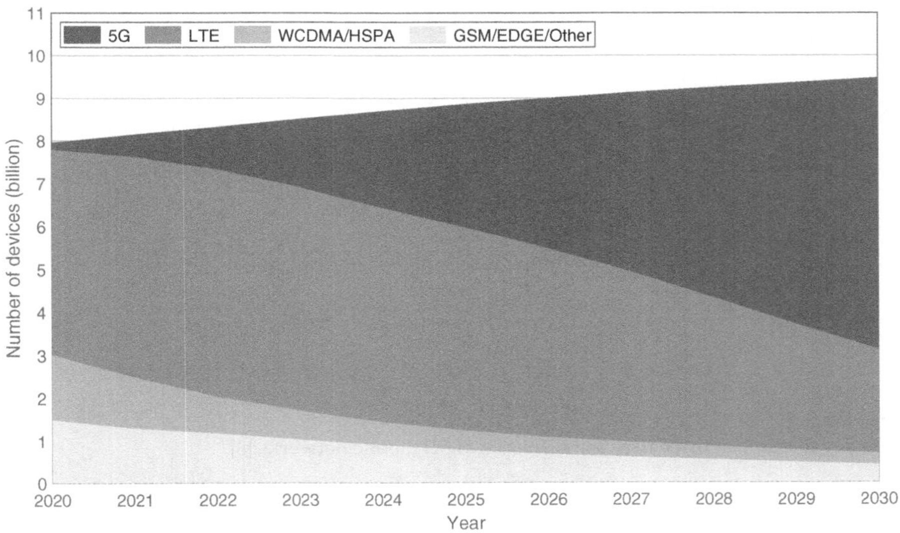

FIGURE 2.2
The projected distribution of mobile network user device subscriptions per different access network technologies in the period from 2020 to 2030 [7]

particularly through the implementation of 5G SA network architecture, will enable the expansion of a range of new services that require high uplink capacity, such as advanced applications for public safety and surveillance.

2.1.2 Key Use Case Scenarios

Global trends are driving the world toward ubiquitous networking, where not only people but also devices communicate with each other, becoming increasingly capable in terms of their ability to communicate without human interaction. As an "everyone to everything" communication paradigm is expected, there is a high demand for network capacity, bandwidth, and coverage, along with the need to ensure network reliability and availability. 5G mobile networks are designed to meet these requirements through the practical application of three key network features (Figure 2.3) defined as: enhanced mobile broadband (eMBB), massive machine-type communications (mMTC), and ultra-reliable low-latency communication (URLLC) [9].

The eMBB use case is designed to provide high capacity and bandwidth, offering users faster and more reliable Internet access, even in areas with a large number of connected devices. mMTC enables the connection of a massive number of devices that require machine-to-machine (M2M) communication, ensuring the efficient operation of IoT systems. With the increasing number of devices and sensors offering M2M communication, mMTC use cases are expected to become more prevalent in the near future. The URLLC use case scenario meets the strict requirements for low latency and

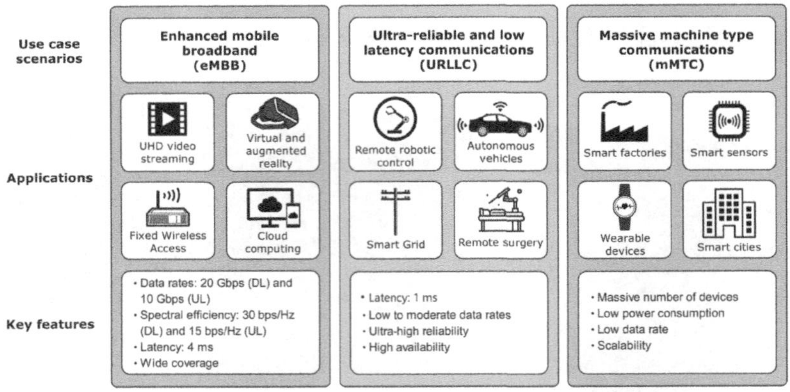

FIGURE 2.3
The key features of three use case scenarios of 5G mobile networks [8]

high reliability, which are crucial for critical applications and real-time communications. These capabilities are essential for emerging applications such as telesurgery, intelligent transportation, and industrial automation [10].

2.1.3 Heterogeneous Network Architecture and Frequency Spectrum of 5G Networks

To optimize coverage, capacity, and EE, 5G networks are deployed as heterogeneous networks (HetNets) which integrate different types of BSs that operate in different operating frequency bands (Figure 2.4). HetNets utilize a combination of outdoor macrocells, supported by macro BSs, to provide extensive signal coverage across large areas, and small cells, such as microcells, picocells, and femtocells, that address needs for high data throughput and enhanced coverage in dense urban areas or indoor environments [11]. Macrocells operate primarily on lower frequency bands, ensuring long-range coverage, while small cells exploit higher frequencies to support data-intensive applications within specific hotspot areas. This multi-layered approach allows HetNets to not only fill coverage gaps but also efficiently manage large capacity demands, providing seamless connectivity and high quality of service (QoS) for users. Additionally, by reducing the distance between users and BSs in densely populated areas, HetNets can also contribute to the overall reduction of BS transmission power, which consequently can have a positive impact on network EE.

The HetNet architecture enables greater flexibility in network design and management, which is critical for 5G networks that must support a wide range of applications, from high-bandwidth video calls to mission-critical, low-latency data transmission applications. The advantage of 5G HetNets architecture is the ability to efficiently distribute resources according to user needs, which enables better utilization of spectrum and energy resources.

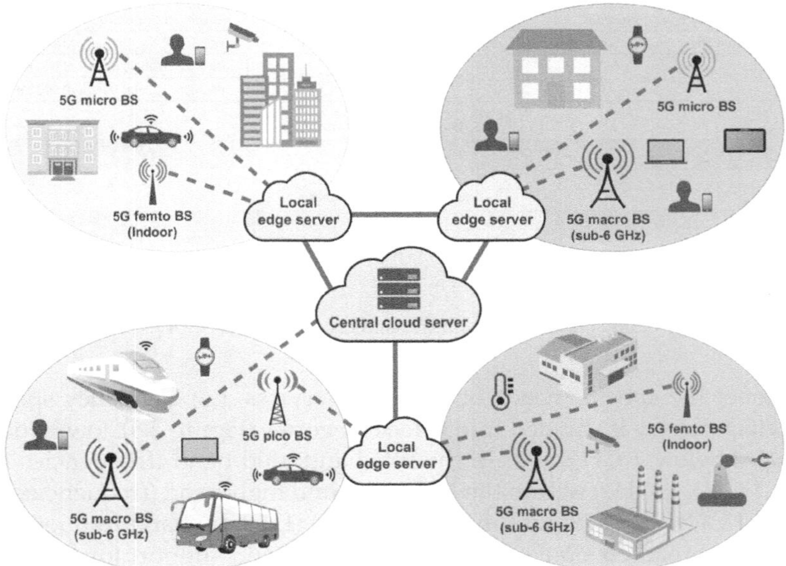

FIGURE 2.4
The 5G HetNet architecture consisting of multiple BS types and edge computing network infrastructure (servers) [12]

To overcome the limitations of the operation in sub-6 GHz frequency bands of mobile networks due to heavy congestion of this frequency spectrum, 5G network standardization enables the usage of millimeter wave (mmWave) frequency bands for the operation of 5G networks. These 5G networks operate at frequencies above 24 GHz, enabling wider frequency channels and significantly higher data rates [13]. However, mmWave frequencies have limitations, which are primarily reflected in the short range of signal transmission, due to high signal losses during signal propagation and penetration. This signal loss is caused by the negative influence of obstacles or weather conditions on the transmission of wireless signals in mmWave frequency bands [13]. Therefore, a heterogeneous network architecture is essential for the deployment of 5G networks, with macro cell BSs providing wide coverage of the 5G network, while small cell BSs complement the network by enabling high capacity in areas with high user density and capacity demands (hotspots).

With the exploitation of the mmWave frequency bands in 5G networks, up to 100 GHz of new spectrum is expected to become available for mobile communications. This represents a significant increase in the available bandwidth of wireless channels compared to the channel bandwidths of earlier mobile network generations utilizing channel carrier frequencies in a spectrum that is below 6 GHz.

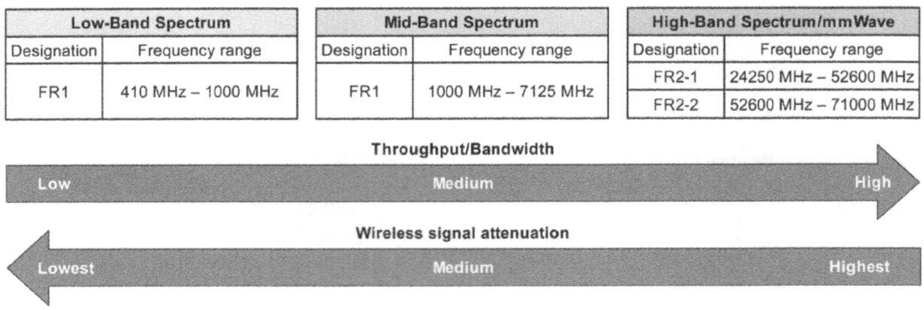

FIGURE 2.5
The 5G frequency band classification and transmission characteristics [14]

To meet the specific needs of different services, the frequency spectrum of 5G technology is divided into three categories (Figure 2.5): low-band (frequencies below 1 GHz) within the FR1 band, mid-band (frequencies in the range 1 GHz–6 GHz) within the FR1 band, and high-band (frequencies in the range 24 GHz–52 GHz) within the FR2 band [15]. Low-band frequencies are crucial for broad coverage, indoor penetration, and support for IoT applications. The mid-band spectrum offers a balance between coverage and capacity, with the expected use of the frequency band in the range of 3.3 GHz–4.2 GHz for most commercial 5G networks [16]. To achieve extremely high data transfer rates and ultra-low latency, which are essential for some specific real-time services in 5G and future 6G networks, operation at mmWave frequency bands is required. Globally allocated mmWave frequency spectrum includes frequencies ranging from 24 GHz to 28 GHz, with plans for additional spectrum allocations in the future (e.g., 37 GHz–50 GHz, 50 GHz–71 GHz) [16].

2.1.4 Energy Efficiency of 5G Networks

The emphasized need for better EE of mobile networks has become a priority in the planning and deployment of 5G networks. This priority arose primarily due to the high energy requirements associated with the dense allocations of a high number of BSs in 5G HetNet architecture, which is necessary to support a large number of users and continuously increasing data traffic volumes [17]. The number and type of 5G BSs required for ensuring network coverage and capacity in a given area directly affects the energy footprint of the mobile network. For example, in highly urban areas, ultra-dense networks with combined macro and small BSs can address high network capacity requirements, but can also significantly contribute to an increase in mobile network energy consumption (EC).

Thus, the number of installed 5G BSs depends on the number and density of user devices and their demand for data capacity and transmission speeds. Variations in user device number, density, and capacity demand in different areas require different 5G network configurations in terms of the number and capacity of 5G BSs. These differences lead to different effects on the EC and consequently on the EE of 5G networks [18]. As more macro and small cell BSs are needed to satisfy coverage and capacity demands, the energy requirements of 5G networks will proportionally increase with the installation of new macro and small cell BSs. Hence, with the transition of mobile network technology from 4G to 5G and future 6G, EE of mobile networks becomes a critical concern, as each new generation of networks requires more energy to keep up with the growing number of users and data traffic volumes [19]. The fact that BSs of some previous mobile network generations (such as third generation [3G]) will eventually be completely removed from the mobile network, will not change this trend of increasing network EC.

Therefore, to meet this versatile need for coverage and capacity in different UD density regions such as dense urban, urban, suburban, or rural, 5G networks are deployed as HetNets. HetNet architecture offers an efficient approach to address various user requirements, but also comes with complex power consumption profiles of HetNets due to the operation of multiple different types of macro and small cell 5G BSs in specific parts of HetNets. Consequently, the balance between the number of deployed macro and small cell BSs, as well as the configuration of the BSs in the network architecture that needs to appropriately accommodate user density and capacity demands, becomes a significant factor in achieving optimal EE in 5G networks [18].

In addition, the integration of network applications that are characterized as data-intensive services further complicates energy management in 5G networks. Serving applications demanding high throughput, real-time data processing, low-latency connectivity, and high reliability contribute to the increase in mobile network energy costs, due to higher 5G network resources that need to be deployed to satisfy such specific application demands.

In response to these growing demands, the move toward more energy-efficient 5G networks is aligned with the wider sustainability goals of the telecommunications industry [20]. Implementing network designs that reduce power consumption without sacrificing network performance and QoS is critical to reducing the environmental impact of large-scale 5G and future 6G network deployments. Future developments of adaptive HetNet configurations, which will include dynamic adapting of usage of RAN and CN resources to real-time traffic requirements, represent promising solutions to further improve mobile network EE. These adaptive network configurations can allow RAN and CN elements to enter low-power states during periods of low traffic intensity, thus reducing overall network power consumption while maintaining the high performance required during periods of higher traffic intensity.

Looking ahead, as 5G networks expand to serve an increasing number of different data-driven applications, the EE of mobile networks will become highly important. Achieving growth of network capabilities in a sustainable manner which will contribute to reducing mobile network energy costs is key to the future of mobile network development. By prioritizing EE in the development of current and future networks, operators not only reduce OPEX but also positively contribute to global efforts dedicated to reducing carbon emissions and promoting green communications [21].

2.2 The Next Evolutionary Phase of 5G Networks

The "5G Advanced" mobile network represents the next phase in the evolution of 5G networks. It is based on the 3rd Generation Partnership Project (3GPP) Release 18 (Rel-18) version of the technical specification, aiming to further enhance the performance, coverage, and EE of existing 5G networks. This technology introduces key innovations that enable mobile operators to provide intelligent network solutions tailored to the specific needs of users, such as autonomous vehicles, industrial automation, and applications of extended reality (XR) [22]. 5G Advanced is often described as the initial stage of development of technologies beyond 5G (B5G), which will continue the evolutionary path toward 6G networks.

Figure 2.6 illustrates the evolutionary path of 5G networks, from the beginning of 5G technology standardization with the Release 15 (Rel-15) specification to the planned introduction of 6G technology as part of the Release 21 (Rel-21) specification. The initial development of 5G networks, designated as 5G Basic in the Rel-15 specification, introduced foundational functionalities such as enhanced eMBB and an initial framework for URLLC use cases. Subsequent development phases, through the Rel-16 and Rel-17 specifications, known as the 5G Evolution, brought extensions in the form

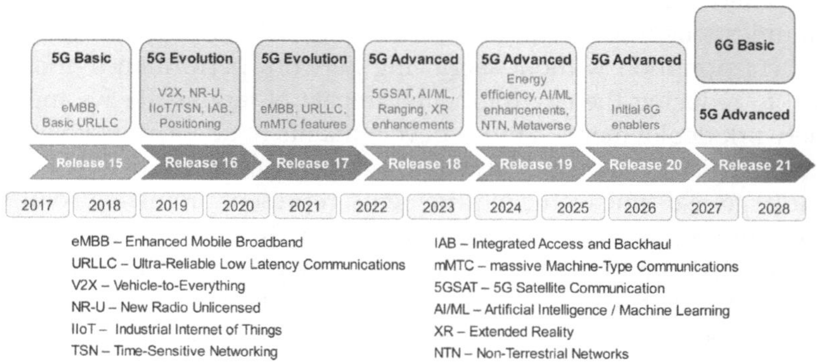

FIGURE 2.6
The evolutionary path of 5G network standardization leading to 5G Advanced and 6G network standardization [22]

of vehicle-to-everything (V2X), industrial IoT (IIoT), and time-sensitive networking (TSN) technologies, as well as additional features for mMTC use cases.

Starting with the Rel-18 specification, 5G technology is entering the 5G Advanced phase (in the period 2022–2028), which introduces further improvements in throughput, capacity, and reliability, in order to meet increasing connectivity and service demands. Finally, the Rel-21 specification, which is scheduled for 2028, will initiate the development of 6G Basic technology, thereby introducing a new era of mobile communications with advanced connectivity options and new paradigms in the use of mobile networks.

One of the most important features of 5G Advanced technology is enhanced network management and improved operating efficiency through the application of artificial intelligence (AI) and machine learning (ML). This technology allows for users' connectivity adaptation to specific services, while simultaneously increasing network EE and expanding support for IoT devices, including those without their own power sources. Furthermore, 5G Advanced allows seamless connectivity between terrestrial and non-terrestrial networks, thus greatly enhancing coverage in rural and difficult-to-reach areas [22].

In addition to these technological advancements, 5G Advanced enables operators to more effectively monetize their investments in 5G networks by establishing new business models and strategic partnerships. It is anticipated that this technology will be essential in the transition to 6G networks, thereby establishing the foundation for future advancements in wireless communications [23].

2.2.1 Technological Innovations within the 5G Advanced Paradigm

The evolutionary phase of 5G technology encompassed by the 5G Advanced paradigm brings significant improvements in the radio part of 5G networks. Table 2.2 shows the key use case scenarios defined in the 5G Advanced networks concept [23]. These scenarios represent the main innovations that define the future development of 5G Advanced technologies as a further step toward the development of 6G networks. Although the implementation of most of these use cases of the 5G Advanced paradigm presented in Table 2.2 will contribute to an increase in network EC, some of them will also have the potential to contribute to the improvement of the 5G network EE.

For example, new use cases of the 5G Advanced concept presented in Table 2.2 include the usage of advanced AI/ML techniques for optimizing radio interface management of 5G BSs. Such techniques will, through optimization of resource allocation and transmit (Tx) power management of 5G and future 6G BSs, reduce the energy required to transmit each bit of data. Thus, the implementation of AI and ML techniques in the 5G Advanced concept will have a key role in optimizing mobile network EE [24].

TABLE 2.2

Key Use Case Scenarios in 5G Advanced Networks [23]

New Use Case Scenarios	Description
AI/ML application in BS wireless (air) interface	Use of AI/ML to improve feedback on channel state information (CSI), manage 5G signal beamforming, and enhance positioning accuracy
Enhancements in positioning and data transmission with NR sidelink technology	Improvements for applications like V2X, public safety, and commercial services, with a focus on positioning integrity for critical services
NR RedCap-related use cases	Support for wearable devices, industrial sensors, video surveillance, and smart grids with reduced device complexity
Enhancements for UAV NR services	Addressing interference issues in uplink and downlink and improving mobility for UAVs
Enhancements for NR multicast and broadcast services	Improvements for public safety services, critical applications, V2X, Internet Protocol Television (IPTV), and IoT applications through multicast and broadcast services
Use case scenarios related to mobile integrated access and backhaul (IAB), and vehicle-mounted relay (VMR) technologies	Expanding coverage and capacity in vehicles, at public events, and during emergencies through mobile IAB networks
Improvements for XR services	Optimization of latency and EE for AR, real-time media, and industrial control applications
Dual Tx/Rx multiple universal subscriber identity module (MUSIM) for connected operation on both universal subscriber identity modules (USIMs)	Ability to simultaneously use voice calls on one USIM while downloading data via the other USIM
Further data collection enhancements for self-organizing networks (SON) and minimization of drive tests (MDT) in 5G NR and evolved non-standalone dual connectivity (EN-DC) configurations	Enhancing data collection and usage for implementing self-organizing network concepts and minimizing drive tests

According to Table 2.2, the 5G Advanced concept considers improvements in NR sidelink positioning that enables direct communication between 5G devices without the mediation of BSs. This technique can improve network EE through accurate user positioning, which reduces retransmissions caused by signal loss or interference. Also, this technique can contribute to minimizing energy waste associated with routing through the CN infrastructure, since managing signaling traffic at BSs and CN nodes will be reduced, which additionally reduces overall network load, enabling BSs to operate in low-power modes more frequently.

Another 5G Advanced use case is the support for 5G NR reduced capability (RedCap, also known as 5G NR-Light) devices that can be implemented in wearables, IIoT, smart cities, automotive, connected healthcare, etc. These devices have moderate throughput, power supply, and performance requirements and thus lack the need for the full capabilities of standard 5G devices. RedCap technology supports 5G Advanced beamforming and massive MIMO (mMIMO) technologies that focus energy precisely on user devices rather than broadcasting signals broadly, thus contributing to an increase in network spectral efficiency (SE), reduction of network energy losses, and optimization of BS power usage. In addition, 5G RedCap technology enables efficient multi-RAN coordination between 5G and networks of previous generations (e.g., 4G) to offload traffic intelligently, thus balancing network load and reducing energy consumption across heterogeneous RANs.

Also, the 5G Advanced concept envisions the usage of AI and edge computing to optimize unmanned aerial vehicle (UAV) flight paths based on network coverage, signal strength, and traffic demand. This enables the creation of custom network slices tailored for UAV services that can ensure efficient allocation of network resources from the 5G BSs placed on UAVs, which avoids overprovisioning of network resources and leads to lower network energy use.

According to Table 2.2, another use case in the 5G Advanced framework is the implementation of NR multicast and broadcast services. Although the introduction of multicast and broadcast services generally represents a significant burden to mobile network EE, the new enhancements in NR multicast and broadcast services under the 5G Advanced framework have the potential to improve the energy efficiency of mobile networks by optimizing how data is transmitted to multiple users. These enhancements enable efficient delivery of the same content (e.g., live video streaming, emergency alerts, software updates) to multiple devices simultaneously, reducing redundant transmissions and lowering overall BS energy consumption. These improvements not only reduce the energy footprint of mobile networks but also support sustainable growth in data-intensive applications like video streaming, IoT updates, and emergency communications.

Supporting new mobile network market segments that rely on data-intensive and consequently energy-demanding applications such as XR applications is also seen as an important use case in the 5G Advanced paradigm (Table 2.2). By leveraging techniques based on edge computing, AI-driven optimization, adaptive streaming, and dedicated network slices, energy usage can be reduced, while the data-intensive demands of XR services can be supported. The advancements in these technologies can ensure that XR can scale sustainably, enabling immersive experiences without compromising network EE. Furthermore, given the expected growth of XR applications, the 3GPP Rel-18 specification introduces standards to reduce data transmission delays and improve the EE of XR devices.

Moreover, the 5G Advanced concept envisions data collection enhancements for self-organizing networks (SON) and minimization of drive tests (MDT) in 5G NR networks. By enhancing data collection and leveraging the SON concept, the 5G Advanced framework can improve mobile network EE by reducing energy wastage through dynamically adjusting the usage of network resources based on traffic patterns and user behavior. Also, MDT in 5G NR eliminates the need for vehicles, equipment, and personnel to perform physical drive tests, significantly reducing operational energy consumption and carbon dioxide emissions. These advancements thus can reduce both the operational and environmental energy footprints of mobile networks, paving the way for more sustainable and efficient connectivity.

Also, use case scenarios of the 5G Advanced concept related to mobile integrated access and backhaul (IAB) networks intelligently combine access and backhaul functionalities within the same radio resources and infrastructure based on real-time traffic demands. This can result in a reduction in network complexity and in operational energy costs. The IAB concept thus removes the need for standalone backhaul links (e.g., fiber or microwave) and exploits small cells and relay nodes for backhaul transmission, which can contribute to reducing energy consumption associated with powering separate backhaul systems. More specifically, fewer network components for data transmission across the network reduces energy overhead for network operations and simplifies network maintenance.

Finally, dual transmit (Tx)/receive (Rx) multiple universal subscriber identity module (MUSIM) technology of the 5G Advanced concept enables connected operation on MUSIMs by intelligently selecting the optimal network for each subscriber identity module (SIM) based on coverage, load, or other metrics that might include network EC. Dual SIM operations can enable better traffic distribution across networks by balancing network load and avoiding energy-intensive operations on overloaded cells, thus reducing the energy burden on overloaded BSs. Additionally, by supporting dual active connections, MUSIM devices can reduce dropped calls or session interruptions [25], which otherwise require energy-intensive reconnections. Also, optimized coordination for dual SIMs can reduce signaling complexity through synchronized updates and periodic signaling across both SIMs, lowering the processing energy requirements for core network elements.

2.2.2 A Transition toward 6G Networks

The 5G Advanced phase of mobile 5G network evolution allows service providers to take advantage of new market opportunities, while opening the possibility for reducing network energy requirements and improving overall network efficiency. With the introduction of the 3GPP Rel-18 specification, key improvements that were introduced expanded the capabilities of 5G networks, including improvements in positioning, data transmission, and support for critical applications such as public safety and V2X communications [23].

Thus, all of the discussed use cases envisioned in the 5G Advanced concept not only represent advances in the 5G technology but also open up new opportunities for business models and innovations across industries. Future 3GPP Rel-19 specifications will further include advanced AI/ML technologies for mobility management and network optimization, which will be crucial for supporting upcoming 6G networks [26].

According to [24], the evolution toward B5G networks will be key to connecting the future society, enabling advanced forms of communication with the integration of digital twins and cyber-physical systems. However, the emphasis will continue to be on HetNet architectures, which combine different types of BSs and frequency bands to ensure needed coverage and capacity, while optimizing network EE.

Considering the future of wireless communications, the transition from 5G to 6G technologies represents an important phase. Therefore, 5G Advanced represents an important step toward 6G technologies, providing the foundation for future innovations in wireless communications. This technology enables mobile operators to increase capacity, improve coverage, and optimize EE, thereby opening up new opportunities in the market.

2.3 Future 6G Technology

Although 5G technology already brings significant improvements in mobile networks, the development of 6G technologies is aimed at providing ultra-high reliability, achieving extremely high data transfer rates, and reducing data transfer latency, with a focus on more energy-efficient systems. These technologies will enable fully autonomous systems, advanced AR capabilities, and the further integration of mobile networks in the realization of smart cities, which will make mobile communication networks even more important for everyday life [27].

The constant progress of mobile communications provides the foundations and conditions for the development of 6G technology, while its direction is shaped by the global needs and requirements of future applications. While 5G enabled transformative advances in data transfer speeds, latency, and device connectivity, 6G is expected to introduce an additional level of integration, intelligence, and adaptability. Thus, the deployment of 6G networks is expected to improve current capacity and create opportunities for new applications, particularly in areas such as AI, AR, sensor networks, and ubiquitous connectivity.

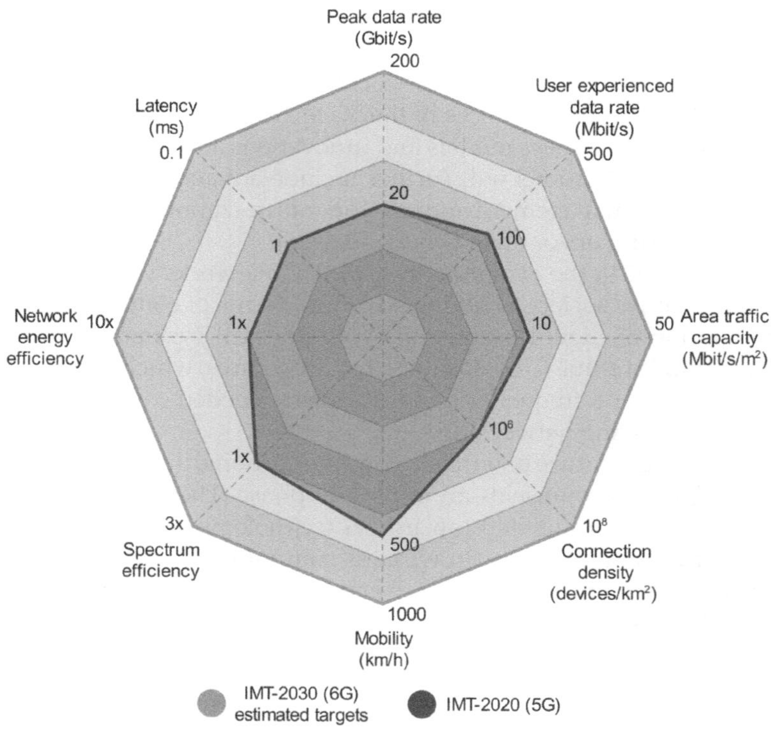

FIGURE 2.7
Comparison of the key features of 5G and 6G technologies [28]

2.3.1 Key Features and Technical Innovations of Future 6G Technology

6G technology is expected to bring significant improvements in all aspects of mobile communications in the next decade. While traditional mobile communications will remain as a key application, new use case scenarios will become increasingly important. As a result, 6G networks will focus on improving the user experience, with an emphasis on high levels of network security, spectral and energy efficiency, intelligence, and customization [1]. This user-centric approach, rather than a purely technology-driven focus, will be a fundamental characteristic of 6G networks.

The chart in Figure 2.7 compares the key features of 5G and 6G technologies based on various criteria, such as EE, intelligence, spectral efficiency, security and privacy, and affordability and customization. The chart clearly shows that the expected features of 6G technology surpass those of 5G in all these aspects. One of the most significant improvements in 6G networks is expected in the areas of security, secrecy, and privacy, which will be crucial for protecting user data while enabling adaptable security protocols to meet different user needs and requirements. Additionally, the greatest emphasis in 6G network development is expected to be on network EE, with the aim of reducing EC and promoting sustainability. Significant advances in network intelligence are also expected, with AI having a key role in smarter and more autonomous network management. Also, affordability and customization will become important features, making 6G networks more cost-effective and adaptable to the unique needs of users and different industries. On the other hand, the least improvement in the development of 6G networks is expected in spectral efficiency, as 5G networks have already made significant progress in this area.

One of the key technological advances in 6G networks is expected in the utilization of terahertz (THz) communications. This represents the next major step in the evolution of wireless networks, overcoming the limitations of 5G technology since 5G is expected to soon reach its limits in terms of the limited availability of spectrum resources [29]. Using the frequency range between 0.1 and 10 THz, communication in these frequency bands enables huge data transfer rates, which are essential for future applications such as ultra-high resolution video streaming, XR, holographic-type telepresence, tactile Internet, and other data-intensive services [30]. On the other hand, THz communication faces significant challenges, including beam direction control due to short signal wavelengths, atmospheric absorption of wireless signals above 1 THz caused by water vapor, and multipath and Doppler effects that require advanced channel estimation to enhance spectral efficiency [31]. Additionally, high sampling frequencies are required to design THz devices, however current technologies are limited with low sampling efficiency and insufficient output power from existing signal sources that prevent full support of THz communications.

Therefore, the key technologies that will enable THz communications include advanced channel characterization modeling, multibeam antenna design, integrated chip development, baseband signal processing, and network resource management [29]. Precise channel modeling is necessary for the support of THz communications due to the specific characteristics of THz signal propagation, such as high signal losses and unique absorption properties. Multibeam antennas provide transmission utilizing high-gain directed wireless signal beams that can efficiently target multiple users or devices simultaneously, compensating for high transmission losses. The development of integrated chips that will support processing in THz communication systems currently presents a challenge, due to the high processing frequencies and component complexity of chips. In addition, advanced baseband signal processing needs to be developed to support efficient modulation and adaptation to highly variable THz channel conditions. Also, network resource management that includes spectrum and EC management of 6G network elements will be essential for optimizing network performance in terms of capacity and connectivity, while ensuring sustainable network deployment and operation. Together, these technologies lay the foundation for the sustainable integration of THz communication systems in the 6G networks.

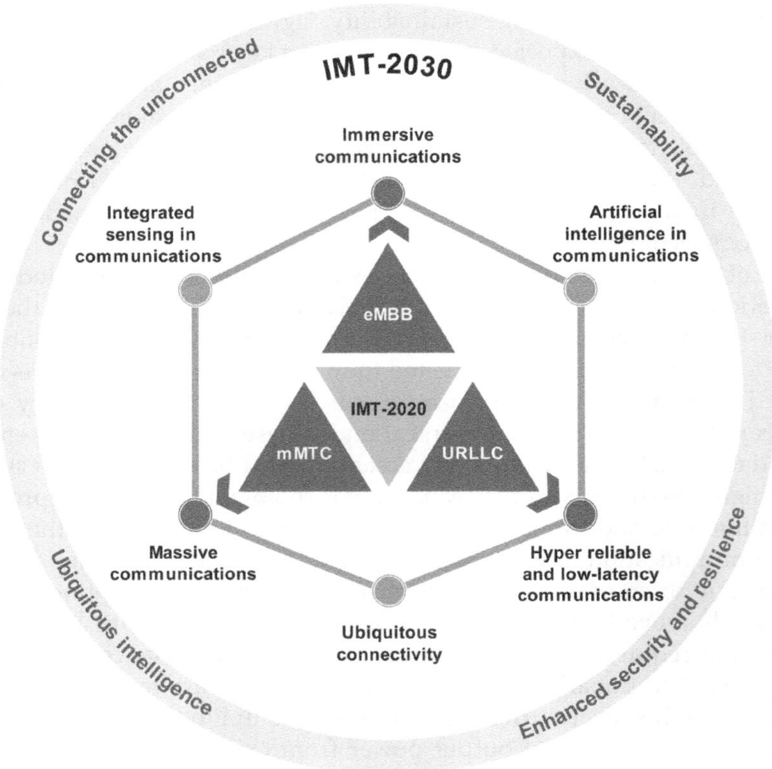

FIGURE 2.8
Evolution of mobile network use cases and services from International Mobile Telecommunications (IMT)-2020 standards toward IMT-2030 vision [28]

2.3.2 Use Case Scenarios and Technological Challenges of Future 6G Networks

The International Mobile Telecommunications – 2030 (IMT-2030) framework of the International Telecommunication Union – Radiocommunication Sector (ITU-R) identifies different capabilities for 6G technology. The IMT-2030 aims to expand existing use case scenarios and enable new, innovative 6G network capabilities, as shown in Figure 2.8 [28]. While IMT-2020 [9] introduces eMBB, URLLC, and mMTC for 5G networks, IMT-2030 introduces new use cases based on AI, sensor technology, and ubiquitous connectivity for 6G networks (Figure 2.8).

One of the key new use cases will be immersive communication, which will enhance eMBB with interactive video experiences, such as XR, remote telepresence, and holographic communications. This technology promises a completely new user experience, enabling the integration of the virtual and real world. Furthermore, hyper-reliable and low-latency communication

(HRLLC) will extend the capabilities of URLLC technology, providing support for applications where minimal latency is critical, such as full industrial automation, critical system control, and telemedicine. Also, massive communication of IoT devices will experience an expansion, enabling the connection of a large number of IoT devices in sectors such as smart cities, transportation, healthcare, and agriculture. This use case scenario will open new possibilities for the automation and optimization of various processes. The inclusion of ubiquitous connectivity in IMT-2030 will be essential for bridging the digital divide, especially in rural and remote areas, providing universal access to digital services. The ubiquitous connectivity concept of 6G networks is expected to expand traditional user connectivity beyond dominantly terrestrial communications to non-terrestrial segments that include UAV, high altitude platforms, and satellite communications, and will also enable integration with underwater communication systems. Furthermore, IMT-2030 will support advanced AI applications and integrated sensing-communication systems, enabling precise and timely data processing for new industrial and everyday applications.

2.3.3 Advanced Non-Orthogonal Multiple Access Technique

Traditional communication techniques based on multiple access, which include time division multiple access (TDMA), frequency division multiple access (FDMA), code division multiple access (CDMA), and orthogonal frequency division multiple access (OFDMA), are encountering challenges when meeting the growing demands for capacity, spectral efficiency, and EE in the last generations of mobile networks. These challenges are dominantly impacted by the large increase in data traffic volumes and the number of connected UDs. To prevent excessive signal interference and guarantee efficient communication, these techniques employ distinct resource allocation strategies, in which each user is allocated a specific resource, such as time, frequency, or orthogonal frequency subchannel. These access methods, however, may result in inefficient spectrum utilization and network capacity in situations of high user traffic intensities and massive user connectivity.

Thus, significant improvements in the implementation of multiple access technologies are expected with the introduction of the 6G network. These improvements are expected to result in a substantial increase in capacity, reduced data transmission latency, and enhanced support for high-speed applications like VR and AR, as well as support for massive connectivity use cases such as the massive Internet of Everything (IoE). In order to optimize resource utilization, novel multiple access methods are being developed. One of the most promising techniques is non-orthogonal multiple access (NOMA) [32], which emerges as an innovative solution that allows simultaneous signal transmission from multiple users within the same frequency resource block. Thus, NOMA is a multiple access technique that allows multiple UDs to share the same time, frequency, and spatial resources simultaneously, distinguishing users by power levels or codes rather than assigning time-divided, frequency-divided, or orthogonal frequency-divided resources as in TDMA-, FDMA-, and OFDM-based systems, respectively. This access method has the potential to significantly improve the efficiency of spectrum usage and network capacity.

The NOMA technique enables more flexible allocation of radio frequency (RF) resources and optimization of network performance, which is a key advantage for future generations of networks such as 6G, where requirements on higher capacity, larger connectivity, and lower latency are much greater [33]. This technology allows for more efficient spectrum usage and reduced interference, making it particularly suitable for future 6G networks that demand high efficiency and flexibility in resource management.

Unlike orthogonal multiple access (OMA) methods which assign different resources to different users to avoid interference, NOMA allows multiple users to share the same RF resource by assigning different power levels to each user. On the transmitter side, NOMA uses superposition coding, where the signals of multiple users are superimposed on the same frequency spectrum, but with different power levels (Figure 2.9).

On the receiver side, NOMA employs the successive interference cancellation (SIC) technique to decode and separate the users' signals within the same orthogonal resource block (Figure 2.9). The SIC process allows the receiver to first decode the signal with the higher power allocation, then subtract this signal from the total received signal, and subsequently decode the lower power signal from the remaining part (Figure 2.9). This process enables the

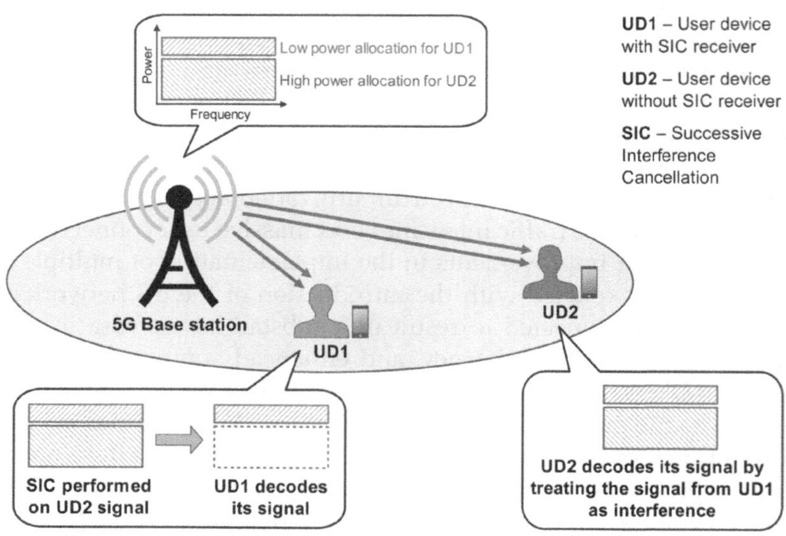

FIGURE 2.9
Application of non-orthogonal multiple access (NOMA) transmission on the user equipment side in the downlink transmission [34]

successful reception of both signals even when they arrive simultaneously, as it eliminates interference from the stronger signal before decoding the weaker one [35]. Thanks to this technique, NOMA can effectively increase network capacity and improve performance for users at the cell edge.

Figure 2.9 illustrates the application of NOMA in the downlink, where the BS uses superposition coding for the simultaneous transmission of signals on the same frequency to UDs in the cell center and at the cell edge [34]. Cell-center UDs equipped with SIC receivers are expected to receive a signal with lower power, as they are closer to the BS and they use the SIC technique to decode the cell-edge user's signal to which a higher signal power has been allocated. After decoding the stronger signal, they reconstruct it and subtract it from the combined signal they initially received. This step leaves only the weaker signal dedicated to cell-center UDs (Figure 2.9), which they can then further process.

On the other hand, cell-edge UDs (Figure 2.9) without SIC receivers decode directly the signal dedicated to these UDs, treating the signal of cell-center UDs as noise. This, however, can result in reduced signal reception quality due to interference. Nevertheless, this approach allows multiple users to share the same frequency resources which increase spectral efficiency and network capacity.

According to research [36], applying the NOMA technique in future radio access networks can lead to a threefold increase in spectral efficiency compared to LTE systems, especially when used with the mMIMO technique and multi-location cooperative techniques. The NOMA technique enables better frequency spectrum utilization and significantly increases mobile network capacity, without the need for additional frequency bands. This makes the NOMA technique particularly suitable for future 6G networks. Such networks require high efficiency and flexibility in resource management, which the NOMA technique provides through its innovative non-orthogonal multiple access approach.

The standardization of the NOMA technique within 3GPP began with the LTE Rel-13 specification through a method known as multi-user superposition transmission (MUST), which focused on downlink (DL) transmissions. Extensions for uplink (UL) transmissions were added in Rel-14 and later specifications [33]. These specifications enable grant-free transmission, reducing latency and improving reliability in 5G network mMTC and URLLC use case scenarios [32]. Further development of the NOMA technique is focused on improving performance and addressing challenges such as the complexity of SIC detection, decoding accuracy, and interference management. Current research specifically targets reducing the complexity and energy consumption of SIC receivers, while enhancing performance in high user density conditions [33].

Within 3GPP, additional applications of the NOMA technique are being considered, including cooperative communications and open-loop transmission in DL. These applications offer further performance improvements in high-mobility conditions and HetNet configurations [33]. Although currently NOMA is not fully standardized in 5G networks, it is being intensively researched for broader applications in 5G Advanced and 6G networks, where challenges like the complexity of signal processing algorithms on the receiver side are being addressed [37]. Further standardization in B5G (6G) networks will depend on additional research and NOMA technique improvements.

Nevertheless, the NOMA technique is considered a critical technology for the development of 6G networks because of its capacity to facilitate massive connectivity, improve spectral and energy efficiency, and provide flexibility in network resource allocation. This makes the NOMA technique particularly well-suited for applications that require low latency, such as VR and AR, and those that involve a large number of devices, such as IoE [33]. Despite its potential, the NOMA technique comes with many challenges, including the requirement for sophisticated algorithms that need to precisely regulate the power levels of wireless signals, manage network resource allocation, and reduce wireless signal interference. The current research aims to address the unique requirements of the 6G network on the NOMA technique. This includes hyper-low latency and extremely high reliability of communication links by utilizing AI to adjust network transmission parameters and optimization algorithms that need to regulate wireless signal power levels, network resource allocation, and signal interference [38, 39].

2.3.4 Energy Efficiency and Sustainability of Future 6G Networks

As wireless communication technologies evolve toward 6G networks, EE becomes a key factor for ensuring network sustainability and long-term operation. The 6G networks are expected to provide significant improvements in data rates, latency, and reliability in comparison with 5G networks, but will also bring vast challenges in terms of optimization of network EC. Nevertheless, the development of 6G networks will inherently take into account network EE and sustainability as key performance indicators.

The need for a trade-off between SE and EE in 6G networks, which represents a long-standing challenge in wireless communications, is envisioned to be alleviated through different approaches implemented on the network and UD side of 6G networks. On the UD side, it is expected that energy harvesting technologies for the power supply of UDs from radio waves, vibrations, or solar sources will have broad practical implementation. This can reduce UDs' dependence on batteries and thus increase UDs' sustainability. Also, the widespread application of intelligent surfaces that optimize radio transmission environments will enable more efficient propagation of RF signals with more optimal use of RF spectrum and energy.

On the network side, a vital concern in the creation of energy-aware network designs is related to the necessity of integrating sophisticated network resource management strategies based on AI algorithms, that will minimize EC while ensuring optimal network performance [40]. The integration of AI and multi-agent systems (MAS), where autonomous agents work together to optimize network operations, will have an important role in improving EE in future mobile networks. These technologies enable smarter and real-time decision-making that can reduce network EC and improve network performance, especially in challenging environments such as IIoT applications [41].

Additionally, 6G networks are anticipated to significantly incorporate "green BSs" with antenna designs that will employ intelligent energy management techniques in signal transmission and reception and renewable energy sources for the power supply of BSs. Moreover, research focused on future 6G networks suggests that adjusting radio resources such as the number of active antenna elements and adapting transmission power levels according to space and time variations of users' traffic could be beneficial for network energy savings. This strategy potentially allows for more energy-efficient use of BS resources, while maintaining the required network performance [42].

In addition to these comprehensive approaches, specific techniques will also play a key role in reducing EC in 6G networks. Development of more energy-efficient hardware components built into 6G network equipment, such as more energy-efficient power amplifiers, frequency/voltage adjustable chipsets, electronic systems on chip (SoC), and implementation of fiber optics for connecting the baseband elements with transmitting and backhauling elements of a radio network will have strong proliferation in 6G networks. Also, one approach to increasing EE in 6G networks is seen in the much larger use of direct device-to-device (D2D) communications, which can optimize EC by reducing the need for data transmission through BSs. Research shows that the application of D2D communication, combined with dynamic UD clustering, significantly improves network efficiency and system adaptability to changing conditions in mobile networks [43]. Also, an important concept that is envisioned as a strong contributor to 6G network EE optimization encompasses the implementation of energy-efficient data centers and cloud computing infrastructure, which also have the potential to substantially reduce the energy footprint of overall telecommunications networks [40].

Therefore, with broad integration into the network operation concepts such as energy harvesting, renewable energy sources, intelligent surfaces, energy-efficient hardware, direct D2D communications, and especially AI-driven radio and data center resource management, 6G networks will set new standards for EE and sustainability in wireless communications. These improvements are aimed at building a network ecosystem that meets future connectivity needs, while reducing its environmental impact and thus paving the way for a more sustainable development of telecommunication networks.

References

1. S. Dang, O. Amin, B. Shihada, and M. S. Alouini, "What should 6G be?," *Nat Electron*, vol. 3, no. 1, 2020. doi: 10.1038/s41928-019-0355-6.
2. Tecore Networks, "5G networks – Evolve to the next generation of wireless." Accessed: Aug. 27, 2024. [Online]. Available: https://tecore.com/5g/
3. T. Hayashi, "Evolved Packet Core (EPC) network equipment for Long Term Evolution (LTE)," *Fujitsu Scientific and Technical Journal*, vol. 48, no. 1, pp. 17–20, 2012.
4. H. Holma, A. Toskala, and T. Nakamura, *5G Technology: 3GPP New Radio*. John Wiley & Sons, 2019. doi: 10.1002/9781119236306.
5. E. Dahlman, S. Parkvall, and J. Sköld, *5G NR: The Next Generation Wireless Access Technology*. 2018. doi: 10.1016/C2017-0-01347-2.
6. G. Liu, Y. Huang, Z. Chen, L. Liu, Q. Wang, and N. Li, "5G deployment: Standalone vs. Non-standalone from the operator perspective," *IEEE Communications Magazine*, vol. 58, no. 11, 2020. doi: 10.1109/MCOM.001.2000230.
7. F. Jejdling, *Ericsson Mobility Report*. Stockholm, Sweden: Telefonaktiebolaget LM Ericsson, Nov. 2024.
8. J. Lorincz, Z. Klarin, and J. Ožegović, "A comprehensive overview of tcp congestion control in 5G networks: Research challenges and future perspectives," 2021. doi: 10.3390/s21134510.
9. ITU-R, "IMT Vision – Framework and overall objectives of the future development of IMT for 2020 and beyond," *Recommendation ITU-R M.2083-0*, pp. 1–21, Sep. 2015.
10. H. Chen *et al.*, "Ultra-reliable low latency cellular networks: Use cases, challenges and approaches," *IEEE Communications Magazine*, vol. 56, no. 12, 2018. doi: 10.1109/MCOM.2018.1701178.
11. M. Ali, S. Mumtaz, S. Qaisar, and M. Naeem, "Smart heterogeneous networks: A 5G paradigm," *Telecommunication Systems*, vol. 66, no. 2, 2017. doi: 10.1007/s11235-017-0291-6.
12. V. Chamola, S. Patra, N. Kumar, and M. Guizani, "FPGA for 5G: Re-configurable hardware for next generation communication," *IEEE Wireless Communications*, vol. 27, no. 3, 2020. doi: 10.1109/MWC.001.1900359.
13. 5G Americas, "Understanding mmWave spectrum for 5G networks – white paper," 2020, Accessed: Aug. 21, 2024. [Online]. Available: https://www.5gamericas.org/wp-content/uploads/2020/12/InDesign-Understanding-mmWave-for-5G-Networks.pdf
14. European Telecommunications Standards Institute ETSI, *TS 138 101-1 (V18.7.0) 5G; NR; User Equipment (UE) radio transmission and reception; Part 1: Range 1 standalone*, Sophia Antipolis, France: ETSI, 2024.
15. GSMA, "5G spectrum - GSMA public policy position," 2022. Accessed: Aug. 23, 2024. [Online]. Available: https://www.gsma.com/connectivity-for-good/spectrum/wp-content/uploads/2022/06/5G-Spectrum-Positions.pdf
16. D. Brenner, "Global 5G spectrum update," 2020. Accessed: Aug. 21, 2024. [Online]. Available: https://www.qualcomm.com/media/documents/files/5g-spectrum-update-for-mipi-alliance.pdf
17. P. Frenger and R. Tano, "More capacity and less power: How 5G NR can reduce network energy consumption," in *IEEE Vehicular Technology Conference*, 2019. doi: 10.1109/VTCSpring.2019.8746600.

18. J. Lorincz, Z. Klarin, and D. Begusic, "Modeling and analysis of data and coverage energy efficiency for different demographic areas in 5G networks," *IEEE Systems Journal*, vol. 16, no. 1, 2021. doi: 10.1109/JSYST.2021.3074100.
19. P. Kaur, R. Garg, and V. Kukreja, "Energy-efficiency schemes for base stations in 5G heterogeneous networks: A systematic literature review," 2023. doi: 10.1007/s11235-023-01037-x.
20. 5G Americas, "Energy efficiency and sustainability in mobile communications networks – white paper," Dec. 2023. Accessed: Nov. 04, 2024. [Online]. Available: https://www.5gamericas.org/wp-content/uploads/2023/12/Energy-Efficiency-and-Sustainability-in-Mobile-Communications-Networks-WP.pdf
21. M. F. Safitra, M. Lubis, A. R. Lubis, and M. I. Alhari, "The need for energy-efficient networks: A review of green communication systems and network architectures," in *Lecture Notes in Networks and Systems*, 2024. doi: 10.1007/978-981-99-7569-3_11.
22. Ericsson, "5G evolution toward 5G advanced: An overview of 3GPP releases 17 and 18," 2021. Accessed: Aug. 21, 2024. [Online]. Available: https://www.ericsson.com/4a92c5/assets/local/reports-papers/ericsson-technology-review/docs/2021/an-overview-of-3gpp-releases-17-and-18.pdf
23. 5G Americas White Paper, "Becoming 5G-advanced: The 3GPP 2025 roadmap," 2022, Accessed: Aug. 21, 2024. [Online]. Available: https://www.5gamericas.org/wp-content/uploads/2022/12/Becoming-5G-Advanced-the-3GPP-2025-Roadmap-InDesign.pdf
24. NEC Corporation, *Beyond 5G / 6G vision whitepaper*, Tokyo, Japan: NEC Corporation, 2023.
25. J. Lorincz, L. Chiaraviglio, and F. Cuomo, "A measurement study of short-time cell outages in mobile cellular networks," *Computer and Communications*, vol. 79, 2016. doi: 10.1016/j.comcom.2015.12.006.
26. Ericsson, "5G advanced: Evolution towards 6G," 2023. Accessed: Aug. 21, 2024. [Online]. Available: https://www.ericsson.com/49ce19/assets/local/reports-papers/white-papers/5g-advanced-evolution-towards-6g-v2.pdf
27. C. Buratti, L. Clavier, and A. F. Molisch, "Chapter 10 – perspectives," in *Inclusive Radio Communications for 5G and Beyond*, C. Oestges and F. Quitin, Eds. Amsterdam, Netherlands: Academic Press, 2021.
28. "ITU-R M.2160-0 (11/2023) M series: Mobile, radiodetermination, amateur and related satellite services framework and overall objectives of the future development of IMT for 2030 and beyond," 2023. Accessed: Aug. 22, 2024. [Online]. Available: https://www.itu.int/dms_pubrec/itu-r/rec/m/R-REC-M.2160-0-202311-I%21%21PDF-E.pdf
29. C. X. Wang, J. Wang, S. Hu, Z. H. Jiang, J. Tao, and F. Yan, "Key technologies in 6G terahertz wireless communication systems: A survey," *IEEE Vehicular Technology Magazine*, vol. 16, no. 4, 2021. doi: 10.1109/MVT.2021.3116420.
30. W. Jiang *et al.*, "Terahertz communications and sensing for 6G and beyond: A comprehensive review," *IEEE Communications Surveys & Tutorials*, vol. 26, no. 4, pp. 2326–2381, 2024. doi: 10.1109/COMST.2024.3385908.
31. P. Sarkar and A. Saha, "Evolution of 6G and terahertz communication," in *Progress in Optical Science and Photonics*, vol. 23, 2023. doi: 10.1007/978-981-99-0228-6_4.
32. 3rd Generation Partnership Project (3GPP), *TR 38.812 study on Non-Orthogonal Multiple Access (NOMA) for NR*, Sophia Antipolis, France: 3GPP, 2018.

33. Y. Chen *et al.*, "Toward the standardization of non-orthogonal multiple access for next generation wireless networks," 2018. doi: 10.1109/MCOM.2018.1700845.
34. A. Benjebbour, A. Li, K. Saito, Y. Saito, Y. Kishiyama, and T. Nakamura, "NOMA: From concept to standardization," in *2015 IEEE Conference on Standards for Communications and Networking, CSCN 2015*, 2016. doi: 10.1109/CSCN.2015.7390414.
35. S. Sen, N. Santhapuri, R. R. Choudhury, and S. Nelakuditi, "Successive interference cancellation: A back-of-the-envelope perspective," in *Proceedings of the 9th ACM Workshop on Hot Topics in Networks, Hotnets-9*, 2010. doi: 10.1145/1868447.1868464.
36. Y. Saito, Y. Kishiyama, A. Benjebbour, T. Nakamura, A. Li, and K. Higuchi, "Non-orthogonal multiple access (NOMA) for cellular future radio access," in *IEEE Vehicular Technology Conference*, 2013. doi: 10.1109/VTCSpring.2013.6692652.
37. O. M. K. Al-Dulaimi, A. M. K. Al-Dulaimi, M. O. Alexandra, and M. K. H. Al-Dulaimi, "Strategy for non-orthogonal multiple access and performance in 5G and 6G networks," *Sensors*, vol. 23, no. 3, 2023. doi: 10.3390/s23031705.
38. L. You, D. Yuan, L. Lei, S. Sun, S. Chatzinotas, and B. Ottersten, "Resource optimization with load coupling in multi-cell NOMA," *IEEE Transactions on Wireless Communications*, vol. 17, no. 7, 2018. doi: 10.1109/TWC.2018.2830386.
39. X. Xu, Y. Liu, X. Mu, Q. Chen, H. Jiang, and Z. Ding, "Artificial Intelligence Enabled NOMA Toward Next Generation Multiple Access," *IEEE Wireless Communications*, vol. 30, no. 1, 2023. doi: 10.1109/MWC.003.2200239.
40. R. Kumar, S. K. Gupta, H. C. Wang, C. S. Kumari, and S. S. V. P. Korlam, "From efficiency to sustainability: Exploring the potential of 6G for a greener future," 2023. doi: 10.3390/su152316387.
41. D. Negi, F. Prakash, and A. Gupta, "Influence of AI and MAS in enhancing energy efficiency of 6G communications," in *2024 International Conference on Communication, Computer Sciences and Engineering (IC3SE)*, 2024, pp. 234–238. doi: 10.1109/IC3SE62002.2024.10593501.
42. D. Laselva, S. Hakimi, M. Lauridsen, B. Khan, D. Kumar, and P. Mogensen, "On the potential of radio adaptations for 6G network energy saving," *2024 Joint European Conference on Networks and Communications & 6G Summit (EuCNC/6G Summit)*, Antwerp, Belgium, 2024, pp. 1157–1162, doi: 10.1109/EuCNC/6GSummit60053.2024.10597015.
43. S. Aneesh and A. N. Shaikh, "A device to device driven approach towards optimizing energy efficiency for 6G networks," *Indonesian Journal of Electrical Engineering and Computer Science*, vol. 34, no. 3, pp. 1682–1689, Jun. 2024. doi: 10.11591/ijeecs.v34.i3.pp1682-1689.

3

Energy Efficiency of Mobile Networks

According to a GSMA report [1], information and communications technology (ICT) is responsible for 1.4% of global greenhouse gas (GHG) emissions and approximately 4% of global energy consumption (EC). As demand for ICT services and devices continues to grow and the process of transition of ICT systems' power supply to renewable energy remains slow, these numbers are expected to grow in the upcoming years. To achieve objectives dedicated to ensuring global environmental sustainability, the ICT sector must substantially decrease its carbon dioxide (CO_2) emissions, which need to be comparable to considerable reductions of CO_2 emissions in the mid-term period predicted in other industry sectors (such as energy production, agriculture, building, transportation, etc.). Although the CO_2 emissions of the ICT sector have stabilized, a coordinated and comprehensive effort toward the reduction of CO_2 emissions is necessary. This includes prompt policy development for reducing CO_2 emissions and improved CO_2 emissions management.

Without such actions, even stable ICT sector CO_2 emissions at 2020 levels would cause the ICT sector's relative share in global CO_2 emissions in the future to surpass one-third, as other sectors reduce their emissions in alignment with the 1.5°C global warming limit scenario [2]. Hence, it is vital for the ICT sector to actively reduce CO_2 emissions and improve its energy efficiency (EE) to prevent a considerable increase in its share of global CO_2 emissions. Key areas dedicated to achieving this goal include the development of energy-efficient networks and data centers, and the integration of renewable energy sources for the power supply of telecommunication systems.

While the ICT sector contributes to global CO_2 emissions, it also enables significant CO_2 emission reductions in other sectors. According to the Global eSustainability Initiative (GeSI) SMARTer2030 report [3], the ICT sector achieves a benefit 9.7 times greater than its own carbon dioxide equivalent (CO_2e) emissions. Therefore, reducing the ICT system's CO_2 emissions can contribute to environmental preservation, while continuing to support economic growth. However, the growing demand for extensive network connectivity and for the transfer of constantly increasing data traffic volumes drive the ongoing expansion of ICT services and the need for the introduction of new and innovative ICTs. This growth, though beneficial for the global economy and society, also leads to increased contribution of the ICT sector to global EC and CO_2 emissions, posing a substantial challenge for the sector.

DOI: 10.1201/9781003615866-3

3.1 Role of Mobile Networks in ICT Energy Efficiency

Mobile networks have one of the key roles in the ICT sector, and the improvement of mobile network EE has become one of the most important challenges in the development of the ICT sector. The growing demand for ever-increasing data volumes and the increasing number of connected devices are putting pressure on keeping mobile networks energy efficient. This makes energy optimization of mobile networks essential in reducing operational costs, increasing sustainability, and achieving global environmental goals of the ICT sector. Research shows that the mobile industry is responsible for about 31% of the total CO_2 emissions of the ICT sector [1]. One way to improve the energy efficiency of the ICT sector is the adoption of 5G technology, which has the potential to greatly reduce EC per unit of transmitted data [4]. However, the overall impact of 5G networks on the improvement of ICT sector EE and reduction in GHG emissions will depend on how they are implemented, the application of advanced network resource management technologies, and the degree of integration of renewable energy sources.

Until the beginning of the 2010s, improving the EE of mobile networks was not a focus of equipment vendors, mobile network operators (MNOs), and academia, since much attention was focused on ensuring quality of service (QoS), capacity, and data transfer speed in mobile networks, as well as maintaining network functionality through fast errors detection and correction, and reducing the number of cell outages [5]. However, by the end of the 2010s, EE had become an issue in the development and deployment of mobile networks of both industry and academia. This issue has become particularly important concerning the necessity of satisfying the demand for the transmission of constantly growing data volumes and ensuring QoS for the increasing number of connected devices. As the world increasingly relies on mobile technologies, optimizing the EC of mobile networks becomes challenging but crucial for reducing MNOs' operational expenditure (OPEX), improving long-term network sustainability, and meeting the global environmental goals set by different regulatory bodies.

Numerous studies have demonstrated that the primary energy consumers in mobile networks are base stations (BSs), and the optimization of the EC of the radio part of mobile networks can result in substantial energy savings [6]. It was shown that the BS contribution to the overall mobile network EC can be reduced without compromising mobile network QoS, through the implementation of different dynamic resource allocation technologies [7]. In addition to technical solutions, it is essential to establish policies and strategies for sustainable BS energy resource management, in order to decrease EC and GHG emissions of mobile networks.

3.2 Definition of Mobile Network Energy Efficiency

In mobile networks, EE refers to the optimization of EC while maintaining a satisfactory level of QoS for users [8]. Due to the demands for transferring growing mobile data volumes and satisfying the need for constantly increasing data transfer speeds in mobile networks, EE is becoming more and more significant in the design and management of modern mobile networks, particularly with the introduction of 5G and future 6G networks.

EE of the part of the mobile network is typically defined as the maximum amount of data that can be transmitted per unit of energy consumed, commonly measured in bits per Joule (bit/J) [9–11]. Optimization of network EE across all layers of the network, from radio access to core network infrastructure and servers in data centers, enables MNOs to decrease network EC while promoting the mobile network's long-term sustainability. Through the application of approaches for improving EE, mobile networks can achieve a balance between high performance and environmental sustainability.

The standardization of 5G networks starts to include measures for improving network EE. For example, informations about EE targets and the metrics used to measure EE of 5G networks are based on the recommendations and specifications outlined in the standard [12]. Key metrics for assessing mobile network energy efficiency include the ratio between the volume of data transmitted in (part of) the mobile network and the energy consumed by the network elements for the transmission of such data volume. An additional EE metric for assessing mobile network energy efficiency is expressed as the ratio between the area covered with the signal of the mobile network and energy consumed by the network elements that are utilized for ensuring this signal coverage.

The general approach to mobile network energy saving considers partial or complete shutdown of network elements (such as BSs) during periods of low or no traffic transmission. In such an approach, it is important to manage the energy states of network elements, which can range from a complete switch-off state, over some of the partial power-saving states (also known as sleep mode states), to a fully operational state that consumes significantly more energy than other two operating states. The transition among states depends on space and time demands for user connectivity or data traffic transfer and management strategies of network resources that are dedicated to improving network EE.

3.3 Energy Efficiency Metrics in the RAN Segment of the Mobile Network

A radio access part of the mobile network (known as RAN) is generally constituted of numerous BS sites that contain BS equipment and BS site infrastructure. Components of typical macro BS site infrastructure include rectifiers, which convert alternating current (AC) to direct current (DC) power supply, backup power supply systems such as batteries (or diesel generators which sometimes can be the main power supply), cooling equipment like air conditioners, and networking and monitoring equipment.

BS equipment consists of radio units, transmission units (such as antenna elements), and baseband units (BBUs). Networking elements on the BS site such as routers, switches, and microwave devices are used for establishing connections with other BSs or to ensure backhaul connectivity of the BS site with the core network, while monitoring elements are used for control of operating, security, or environmental state of the BS site. Data are transmitted wirelessly between BSs and user devices (UDs) via the air antenna interface of the BSs and UDs. Between the BS site and core network data are transmitted mostly using fiber-optic and copper-based cables or wirelessly using microwave links, and in rare cases using satellite links [13]. Moreover, the BS sites containing small (cell) BSs can have reduced BS site infrastructure compared to macro BS sites whereas, for example, pico and femto BS sites can contain only BS equipment with integrated power supply, natural air cooling, and networking functionalities.

The EE of the BS site is influenced by multiple factors that determine how effectively energy is utilized across different BS site components. These factors include the power supply efficiency of the BS site infrastructure, the operational performance efficiency of the BS, and the radio transmission efficiency of the BS radio interface [13]. Figure 3.1 shows the energy flow through the main components of the macro BS site, starting with the AC input from the power grid. The energy is then converted to DC and delivered to the BS, where it's transformed into the wireless signal power supplied to the antenna, and finally transmitted through the radio air interface to reach the UDs. The energy efficiency of the BS site infrastructure (EE_S), or site EE according to Figure 3.1, is calculated by comparing the amount of DC power supply input to the BS (P_{BS}) with the amount of AC power supplied to the

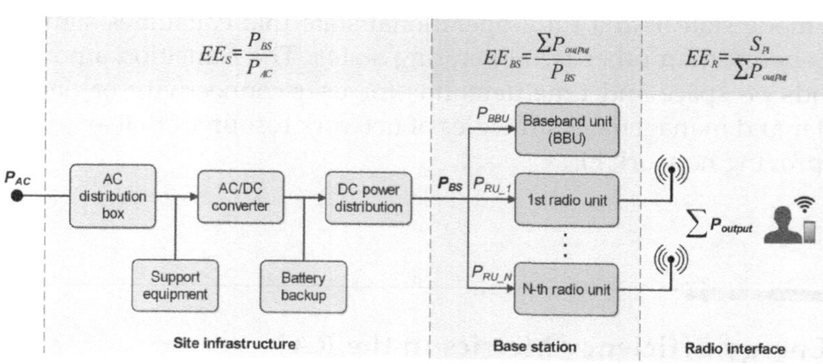

FIGURE 3.1
Energy efficiency (EE) calculation of the base station site [13]

entire BS site (P_{AC}). This metric is the inverse of Power Usage Effectiveness (P_{UE}), which is commonly used for expressing the EE of data centers (DCs).

The operational performance efficiency of the BS (EE_{BS}) is determined by dividing the amount of the transmit (Tx) power supplied to the antenna(s) (P_{output}) with the amount of the overall DC input power of the BS (P_{BS}). Additionally, the radio transmission efficiency of the BS radio interface (EE_R) is measured by comparing the outputs of the BS, such as the volume of transmitted data, the size of the BS coverage area, or the number of served subscribers (S_{pi}) with the Tx power delivered to the antenna (P_{output}). This metric reflects how efficiently the BS utilizes the Tx power of the wireless signal to provide services over the air interface to UDs.

Therefore, the overall EE of the BS site can be represented by the following expression [13]:

$$EP = \frac{P_{BS}}{P_{AC}} \times \frac{\sum P_{output}}{P_{BS}} \times \frac{S_{pi}}{\sum P_{output}} = \frac{S_{pi}}{P_{AC}} \tag{3.1}$$

where:

EP is energy performance of the entire BS site,

P_{BS} is direct current (DC) power input of the BS,

P_{AC} is alternating current (AC) power input of the BS site,

$\sum P_{output}$ is total Tx output power of all BS antennas located at the BS site, and

S_{pi} is service provided by the BS(s) (e.g., delivered bits or number of subscribers).

This ratio presented in (3.1) over a given period of time is expressed in megabits per kilowatt-hour (Mbit/kWh) or megabits per Joule (Mbit/J).

3.4 Standardized Mobile Network Energy Efficiency Metrics

Historically, MNOs invest in the development of new mobile network infrastructure with the main effort related to prioritizing the improvement of network throughput, coverage, reliability, and latency in data transfer. Due to economic and environmental reasons explained in Chapter 2, the development of next-generation mobile networks is now significantly influenced with the need for improving network EE. Thus, MNOs have started to regard EE as a key performance indicator (KPI) in network planning and operation, thereby balancing environmental, financial, and operational objectives while simultaneously ensuring QoS in mobile networks [14].

3.4.1 Network Segmentation for Energy Efficiency Assessment

The most precise assessment of network EE can be done based on detailed knowledge of the EC of each network segment that can be obtained by performing detailed EC measurements. However, a complete and detailed measurement of EC simultaneously for the entire network within one country or for the entire mobile operator is extremely challenging and often impossible. Therefore, in the process of network EE assessment, the total network is commonly segmented into a smaller number of networks of limited size known as "sub-networks" or UD-density areas [9]. Specific characteristics define each UD-density area (sub-network), which for example include capacity-demanding sub-networks representing urban and dense urban areas, sub-urban sub-networks with high coverage and capacity requirements, or rural sub-networks, which are usually less capacity-demanding but with the need to provide mobile network signal coverage to wider areas.

The extent and scale of sub-networks are determined by a variety of criteria, including demographic, geographical, and topological boundaries of sub-networks. Additionally, the classification and analysis of sub-networks can be significantly influenced by a variety of factors, including the level of mobile network penetration, climate zones, and the dominant type of data traffic transferred in specific sub-networks. However, including these factors in the classification of sub-networks can enable a more comprehensive assessment of network EE. Also, this network segmentation on sub-networks allows for the extrapolation of results at the overall network or country level, thereby facilitating a more precise analysis of EE on the level of the whole network.

Depending on the actual practical implementation, the sub-networks generally contain heterogeneous types of BS sites and corresponding BSs which might differ in terms of size (macro or small cell BSs), number of BSs (single or multi-BS sites), BS capacity (single-input single-output [SISO], multiple-input multiple-output [MIMO] and massive MIMO [mMIMO]), operating frequency (single or multi-radio BSs), and technology (2G/3G/4G/5G and B5G).

Moreover, this heterogeneity of BS sites is further complicated with the differences in the implementation of the backhaul part of the mobile network, which enables the connection of BS sites with the core network (CN). Figure 3.2 illustrates an example of the traditional and modern cloud-based mobile network architecture. On the left side of Figure 3.2, a conventional backhaul architecture is shown, in which the central cloud in the CN is connected to the BS site equipment through backhaul connections. However, the modern backhaul mobile network architecture (presented on the right side of Figure 3.2) is based on cloud-based backhaul architecture for connecting heterogeneous BS sites with the CN and central cloud. In this backhaul network architecture, different BS sites, unlicensed local small BSs operating in unlicensed bands (e.g., 2.4 GHz, 5 GHz, and 6 GHz) for localized, high-density

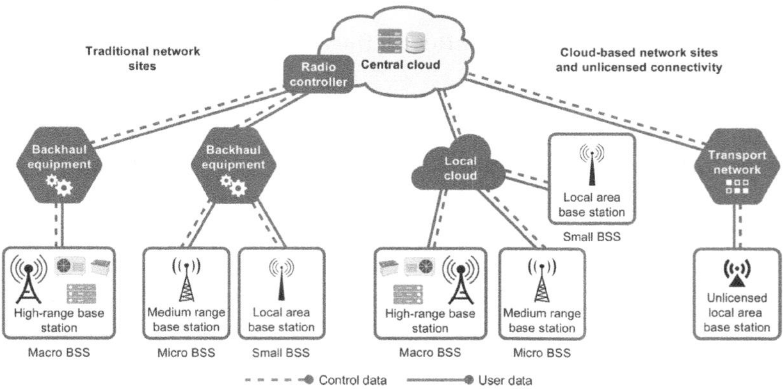

FIGURE 3.2
Comparison of traditional and cloud-based mobile network architectures [9]

deployments dedicated to enhancing network capacity, reducing congestion, or creating private 5G networks, and transport systems are integrated to allow more flexible data flow management over local clouds. These local clouds are used to offload data processing at a central cloud (Figure 3.2). Such modern cloud-based mobile network architecture distributes and balances data transfer through the implementation of the edge computing concept, which reduces network latency and enhances the efficiency, scalability, and performance of the mobile network. Besides these advantages, such complexity and versatility of the backhaul mobile network architecture further complicate the mobile network EC assessment and analyses of network EE.

3.4.2 Parameters for Mobile Network Energy Efficiency Assessment

The sub-network parameters such as overall transferred data volume (DV), coverage area (CoA), and energy consumption are crucial for calculating the EE metrics of specific sub-network parts of a mobile network. These parameters are defined as key parameters for expressing network EE in standards of the European Telecommunications Standards Institute (ETSI) [9], the 3GPP [10], and the International Telecommunication Union – Telecommunication sector (ITU-T) [11].

The overall transferred DV is measured in bits and represents the DV transferred during a given period through a given part (sub-network) of mobile network equipment. It includes data flow in both directions, from the UD (uplink) and toward the UD (downlink). Therefore, the total DV can be expressed by the following expression [9–11]:

$$DV_{MN} = \sum_{i,k} DV_{BS_{i,k}} \tag{3.2}$$

where:

DV is the data volume transmitted in the analyzed sub-network during the measurement period T,

i is the index representing the BS site locations in the analyzed sub-network, and

k is the index representing the number of BSs at the i-th BS site location.

The coverage area of a mobile network refers to an area (commonly expressed in km²) where the sub-network part of the mobile network provides services and has an appropriate level of signal coverage. The coverage area of a mobile network can be expressed through three key parameters [9]. The first parameter is the total geographic area of a country (CoA_geo), which indicates the total area where the mobile operator provides service according to the licensing agreement. The second parameter is the designated coverage area (CoA_des), which indicates the area covered by the mobile network signal according to the network plan that also includes indoor locations. This area significantly depends on the BS Tx power level, the conditions for the propagation of the 5G signal, and the mobile network planning models. The third parameter is the coverage area quality factor (CoA_Qdes), which estimates an actual network coverage area within a given space based on real network performance metrics (such as the number of dropped calls, network efficiency, and QoS). Altogether, these coverage area parameters facilitate in-depth analysis of network coverage.

The EC of sub-network refers to the total amount of energy used by that part of a mobile network during a defined period T [9]. Regarding the EE of a sub-network, it is possible to measure EC at the level of BSs, BS site infrastructure (SI), backhaul network elements (BH), and control nodes in the CN. The network EC is expressed in watt-hours (Wh) and is measured for each individual sub-network component during the measurement period T. The total EC of the sub-network (EC_{MN}) can be represented by the following expression [9–11]:

$$EC_{MN} = \sum_i \left(\sum_k EC_{BS_{i,k}} + EC_{SI_i} \right) + \sum_m EC_{cells} + \sum_j EC_{BH_j} + \sum_l EC_{RC_l} + \sum_l EC_{CC_l} + \sum_l EC_{LC_l} \tag{3.3}$$

where:

$EC_{BS_{i,k}}$ represents EC of the k-th BS at the i-th BS site location,

EC_{SI_i} represents EC of the BS site infrastructure at the i-th BS site location,

EC_{cells} represents EC of small BSs in the mobile network,

EC_{BH_j} represents EC of the j-th backhauling link that provides connection of the BSs to CN,

EC_{RC_l} represents EC of the l-th control node in the CN part of the analyzed sub-network,

EC_{CC_l} represents EC of the *l*-th central cloud (Figure 3.2),

EC_{LC_l} represents EC of the *l*-th local cloud (Figure 3.2),

i is the index representing the number of BS site locations,

j is the index representing the number of backhauling equipment connected to the *i*-th BS site location,

k is the index representing the number of BSs at the *i*-th BS site location,

l is the index representing the control nodes in the mobile network, and

m is the index representing the number of small BSs in the mobile network.

Integrated measurement systems that are incorporated in the software of modern BSs or external sensors can be employed to monitor the EC of equipment belonging to a specific sub-network. These systems can collect EC data from various network components, including BSs, routers, and other BS site infrastructure elements (power supply, batteries, etc.). The collected data can be used to conduct analysis aimed at identifying key sources of energy consumption in order to develop strategies to optimize EE.

3.4.3 Assessment of Mobile Networks' Energy Efficiency

Two standardized metrics are used for assessing the EE of mobile networks as defined in [9–11]. These metrics are mobile network data energy efficiency ($EE_{MN,DV}$) and mobile network coverage area energy efficiency ($EE_{MN,CoA}$). The data energy efficiency of the sub-network ($EE_{MN,DV}$) is defined as the ratio between the amount of overall DV transmitted in the sub-network and the energy consumed by active network devices in the sub-network during the time period *T*. It is expressed in bits per Joule (bit/J) and is calculated using the following expression [9–11]:

$$EE_{MN,DV} = \frac{DV_{MN}}{EC_{MN}} \tag{3.4}$$

where:

DV_{MN} represents the amount of transmitted data in the analyzed sub-network, and

EC_{MN} represents the total EC of the analyzed sub-network.

Coverage area EE ($EE_{MN,CoA}$) is an EE metric defined as the ratio between the coverage area of the sub-network and the annual EC of the analyzed sub-network. This metric is especially useful for sub-networks transferring low data traffic volumes, such as those in rural or sparsely populated areas, and is expressed in square meters per Joule (m²/J). It is calculated using the following expression [9–11]:

$$EE_{MN,CoA} = \frac{CoA\_des_{MN}}{EC_{MN}} \tag{3.5}$$

where:

CoA_des_{MN} represents the designated coverage area of the analyzed sub-network, and

EC_{MN} represents the annual EC of the analyzed sub-network.

Since accurate measurement of EC and the amount of data transferred across different sub-network segments is required for expressing data and coverage EE, assessing data and coverage EE metrics in practice can be quite challenging. Furthermore, the accuracy of the measurements can be challenged by changes in the climate conditions (temperature, humidity, etc.) and the volume of network traffic that occurs during different seasons of a year and periods of a day or week. Despite these challenges, data ($EE_{MN,DV}$) and coverage ($EE_{MN,CoA}$) EE represent critical EE metrics for MNOs and constitute benchmarks for ensuring energy sustainability and EE optimization of mobile network performance.

3.5 Key Factors of Mobile Network Energy Efficiency

The EE of mobile networks depends on several key factors [15] that include network spectral efficiency (SE), traffic variations, latency, interference, and network topology. These factors are essential for understanding the potential for advancement of EE of mobile networks and have great importance in future research dedicated to advancement of EE of mobile communication networks.

3.5.1 Impact of Spectral Efficiency on Network Energy Efficiency

Spectral efficiency refers to the ability of a wireless communication system to transmit as much information as possible over a given frequency spectrum. It is a key parameter in the research and development of wireless networks, especially due to the constant demand for higher data transmission capacity. SE is measured as the amount of data that can be transmitted per unit of the frequency spectrum shared among multiple BS cells in the cell cluster, and it is given by the following expression [16]:

$$\eta \approx \frac{R/B}{K}[\text{bit/s/Hz/cell}] \tag{3.6}$$

where:

R represents the channel capacity or achievable data transmission rate,

B represents the wireless channel bandwidth, and

K represents the number of BS cells in the cell cluster.

A larger cluster size (K) reduces spectral efficiency per cell because the same bandwidth is shared among more cells, while a smaller cluster size allows higher spectral efficiency per cell but can increase co-channel interference. Thus, the relation (3.6) shows the trade-off between maximizing SE per cell and minimizing interference in multi-cell networks. For mobile networks aiming for high SE, smaller cell cluster sizes are preferred but require advanced interference management techniques like beamforming or dynamic frequency allocation.

To have an energy-efficient RAN, both the SE and EE need to be maximized and these two parameters must be balanced [17]. However, the SE and EE are reversely proportional since increases in SE often result in higher mobile network power consumption and in a reduction of network EE [18]. Hence, finding an optimal trade-off between SE and EE is essential for every generation of mobile networks.

The SE measures how well the available frequency spectrum is used to transmit data and can be improved by methods and technologies that maximize the utilization of the same frequency band. These methods include maximizing the data carried per Hz of spectrum by employing advanced modulation and coding techniques and adjusting BS cell Tx power to reduce interference on shared channels [19, 20]. By utilizing precise spectrum sensing, cognitive radio technology also contributes to the enhancement of both SE and EE [17, 21–23]. Heterogeneous wireless networks contribute to SE and EE improvements as well, by using techniques such as complete BS or BS component sleep modes, cell-size reduction through BS Tx power scaling, and green BS architectures that include BS having renewable power supplies, compact and modular designs, advanced cooling solutions (such as air cooling), and smart antenna systems [17]. By minimizing interference and improving signal quality, these strategies collectively support higher data transmission rates within the same frequency band, advancing SE in mobile networks.

As an example, Table 3.1 presents the downlink spectral efficiency for the LTE (4G) and 5G BS using the same frequency spectrum, bandwidth, and antenna configuration. The SE performance differs significantly between 4G and 5G networks, with 5G showing substantial improvements in SE (Table 3.1). Compared to the LTE system using 2 × 2 MIMO antennas having SE of 1.7 bit/s/Hz/cell, 5G can achieve 30–65% higher SE. Furthermore, 4G achieves

TABLE 3.1

Expected Downlink Spectral Efficiency [24]

Frequency Spectrum	Bandwidth	Antennas	LTE (bps/Hz/cell)	5G (bps/Hz/cell)
Sub-1-GHz	10 MHz FDD	2 × 2 MIMO	1.7	2.2
2 GHz	20 MHz FDD	4 × 4 MIMO	2.5	3.3
3.5 GHz	100 MHz TDD	mMIMO	6.3	10.4

50% better results when using a 4 × 4 MIMO configuration compared to a 2 × 2 MIMO configuration, while mMIMO results in a 150% improvement for 4G SE. Ultimately, 5G can achieve an SE that is up to six times higher than LTE when used in combination with mMIMO technology and 100 MHz carriers [24]. Thus, the most significant enhancements are observed at higher frequencies, where advanced antenna configurations, such as mMIMO and wider bandwidths, are implemented. In general, the transition to 5G provides substantial SE gains, establishing a robust foundation for accommodating the high data requirements of the future 6G networks. The expectation for the future 6G networks related to the SE is for the peak, system, and user-perceived SE in the ranges of 60–100 bit/s/Hz/cell, 10–15 bit/s/Hz/cell, and 5–15 bit/s/Hz/cell. Thus, the future 6G network improvements in SE will be essential in meeting the unprecedented demands of next-generation applications and services, and will also contribute to an increase in 6G network EC.

3.5.2 Impact of Traffic Variations on Network Energy Efficiency

The lifestyle of the users and specific use cases of mobile networks significantly impact the variations in mobile network traffic that can have repetitive patterns correlated with daily, weekly, and seasonal user activities. Also, variations in mobile network traffic can be different among different areas such as residential, business, industrial, touristic, etc. Additionally, mobile data traffic in some locations is sometimes impacted by special event-driven traffic variations that include sports, music, or emergency-driven events [15].

In the present day, the traffic patterns of the most popular applications, such as messaging services, communication applications, and social media platforms, are distinct from traditional mobile traffic [25]. Based on data from 2023, mobile social networking and messaging applications make a substantial contribution to the total volume of mobile traffic [26]. Traffic is usually highest during specific times of the day, such as in the morning (7:00 AM–9:00 AM) due to increased usage of navigation, streaming, and messaging apps and evening hours (6:00 PM–10:00 PM) due to high demand for video streaming, social media, and gaming. Late-night and early-morning hours (12:00 AM–6:00 AM) are mostly off-peak hours with the least traffic volumes, which are dominated by background activities like application updates and software downloads. However, precise traffic patterns are determined by location of the users. Even though urban areas experience a higher volume of traffic in general, analysis of network traffic indicates that suburban areas experience the highest volume of individual traffic during peak periods [27]. The complexity of traffic variations in mobile networks presents a significant challenge to MNOs in planning and deploying mobile network infrastructure, which is mostly capacitated to satisfy peak traffic volumes. The consequences of this approach form one of the most important reasons why mobile networks are huge energy consumers and why MNOs as business subjects have been, in almost every country, among the highest individual energy consumers.

In response to these traffic fluctuations, improving the EE of mobile networks is based on the dynamic adjustment of network resources to ensure optimal performance and reduce EC. This encompasses the implementation of advanced technologies that enable energy-efficient management of network resources, while guaranteeing a high QoS under conditions of traffic fluctuations [28, 29].

3.5.3 Impact of Latency on Network Energy Efficiency

Latency in mobile networks is the time delay that passes between sending and receiving data, and has a direct impact on the EE of mobile networks. High latency increases EC at both the network and UD ends, due to inefficient data handling and prolonged active operating states of network elements. High latency delays data transmission and forces retransmissions in mobile networks when time-sensitive application counters expire. UD and network elements stay in an active operating mode longer while waiting for delayed data caused by latency, which increases the EC of network elements and the battery depletion of UDs. High latency disrupts mobile network resource scheduling, leading to the over-provisioning of network resources to compensate for delays, which in turn increases the EC of the mobile network. For example, high latency in the backhaul and core part of the mobile network causes inefficient data processing and longer data routing that contribute to an increase in network EC. Thus, minimizing data transmission latency in mobile networks is essential not only for enhancing QoS but also for improving overall network performance and EE.

Although the implementation of advanced technologies that enable energy-efficient management of network resources is necessary for improving network EE, this implementation can have negative implications in terms of increasing latency in mobile networks. For example, MNOs may experience the problem of latency in the data transfer when implementing approaches for improving network EE that are based on putting RAN equipment in sleep (shutdown or low power consumption) states during periods lacking or having low network traffic. When elements or entire BSs are put into sleep mode, the time required to reactivate them when needed can result in increased data transfer latency. To contribute to the reduction of transition time from some power-saving state to the full operation state of complete BSs, different individual BS elements are considered as potential candidates for inclusion in energy-efficient management of network resources such as BS frequency resource blocks, channel carriers, power amplifiers, baseband units, or antenna elements.

Also, certain network attributes can be modified to mitigate the negative effects of power-saving (also known as sleep) techniques on network latency and performance. The guard interval and hysteresis time are two examples of these attributes [15, 30–32]. The hysteresis time is the time period used to prevent frequent state changes of a BS between active and sleep states, while the guard interval is the predefined time period during which the network can tolerate latency, ensure network stability, and maintain quality of service before reactivating a BS. By modifying these parameters, it is possible to mitigate the negative consequences of latency while at the same time achieving network energy savings. By optimizing the guard interval and hysteresis time, it is possible to achieve a balance between BS operational and EE performance, thereby reducing latency without sacrificing energy savings.

3.5.4 Impact of Interference on Network Energy Efficiency

In the concept of developing mobile networks as heterogeneous networks (HetNets), the deployment of small BSs with limited coverage (also known as small cell BSs) can cause significant inter-cell interference. When small cells (such as micro, pico, and femto cells) are deployed in HetNets, interference management becomes critical to ensure optimal performance and EE of the HetNets [33].

Interference is the phenomenon of disrupting wireless signal transmission which occurs when multiple neighboring BSs and corresponding cells use the same or overlapping frequencies. This problem is especially present in urban areas having dense concentrations of UDs, for serving which a dense allocation of BSs and corresponding frequency resource blocks must take place, thus increasing the probability of signal interference. Therefore, interference can have a negative impact on the user experience through a reduction of network throughput, increased latency, and decreased overall network QoS.

According to [34], the complexity of interference management in 5G and future 6G networks will continue to grow, since the introduction of new techniques such as transmission using non-orthogonal multiple access (NOMA), mMIMO, and dynamic time division duplex (TDD) techniques introduces new forms of interference. These technologies, while delivering significant improvements in mobile network capacity and coverage, also create challenges in interference management, especially in HetNet environments having a dense allocation of UDs.

As interference management becomes increasingly complex with the implementation of densely deployed 5G network BSs, advanced interference control strategies must be developed and implemented in order to achieve a balance between network performance and EE. This is because interference in mobile networks negatively affects network EE by causing wireless signal degradation, higher network processing overhead, and increased Tx power requirements of network elements. More specifically, when interference disrupts signal quality, BSs and UDs compensate for this disrupted signal quality by increasing Tx power to maintain connectivity. This leads to higher instantaneous power consumption for both, the network infrastructure elements and UDs, which consequently results in consuming more energy. Also, interference lowers the signal-to-noise ratio (SNR) at the position of UDs, which results in reduced data transmission efficiency and can cause increased data retransmissions. Since data retransmissions waste mobile network resources, they lead to increased energy consumption of network elements and UDs. Also, high interference leads to an increase in the number of active network elements that need to be active for compensating network performance losses caused by interference. This results in increased energy consumption across the entire mobile network parts. Thus, managing interference effectively is crucial for reducing mobile network EC and improving overall mobile network performance.

3.5.5 Impact of Network Topology on Network Energy Efficiency

The EE of mobile networks is significantly affected by the network topology, which is defined by the spatial arrangement and connections between BSs and UDs [15]. Since BSs must be placed relative to the location of UDs to maintain the appropriate signal level and coverage, irregularity and dispersion in their placement contribute to an increase in network EC. Intensive changes in the position of users and in the layout of BSs due to the activation and deactivation of power-saving modes of the BS elements create challenges related to the emergence of wireless channel conditions having complex stochastic geometry. Thus, network performance is influenced by complex wireless channels that exhibit varying degrees of interference, signal reflection, refraction, and absorption. In order to reduce the impact of dynamic changes in network topology on network EC, it is essential to understand and analyze these challenges, which require additional research to identify optimal network topologies that can maximize network EE in light of the increased dynamic heterogeneity of 5G networks and the expected even larger dynamic heterogeneity of future 6G networks [35]. By optimizing the placement of BSs in mobile networks, it is possible to enhance network coverage and thus reduce the overall Tx power requirements of BSs, which can result in improved network EE.

3.6 Energy Efficiency of Base Stations

Since the highest number of individual network elements in the RAN of MNOs is related to BSs, the BSs stand out as the key RAN component and the highest energy consumer in the RAN segment of the mobile network [36]. The large number of BSs in RAN contributes significantly to the overall EC of mobile networks, thus reflecting their key role in the energy profile of the entire mobile network. A study in [37] highlights that the RAN is the

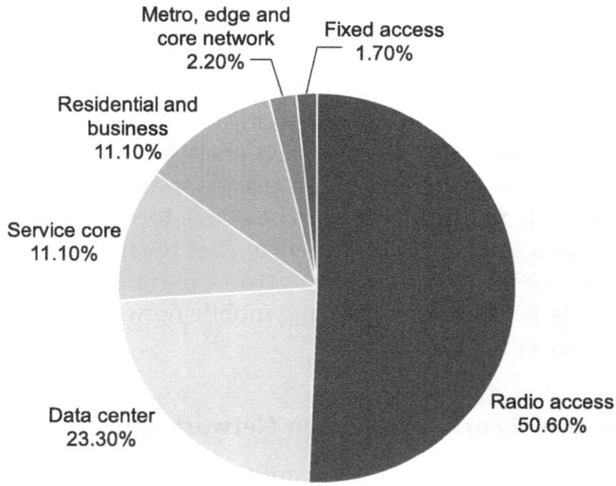

FIGURE 3.3
Estimated distribution of mobile network elements energy consumption in 2025 [38]

main energy consumer in mobile networks, accounting for 57% of the total network EC.

According to [38], due to the expansion of 5G mobile networks, in 2025, the RAN will remain the largest energy consumer in the mobile network, accounting for approximately 50.6% (Figure 3.3) of the total EC of an MNO that offers both fixed and mobile services. Therefore, efforts dedicated to improving the RAN EE are found to have the highest contribution to the improvement of mobile network EE.

A conventional BS consists of different components, as shown in Figure 3.4, that include the antenna with remote radio heads/units (RRUs/RRHs) or the radio unit with radio frequency (RF) chain(s) including duplexer and low-noise amplifiers, the wireless signal power amplifier (PA), the BBUs, the DC-DC power supply, and the main power supply [39]. When evaluating the EC of BSs, the transmission power consumption, computation power consumption, and additional power consumption are considered to be the three

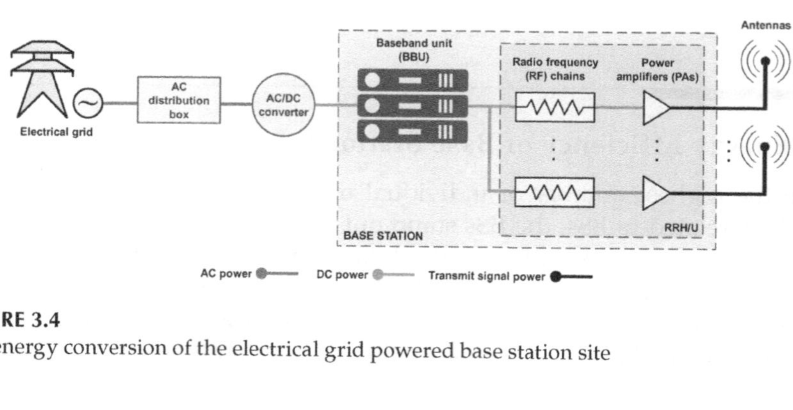

FIGURE 3.4
The energy conversion of the electrical grid powered base station site

main components that contribute to the overall instantaneous BS power consumption [40].

Transmission power refers to the power consumed by the RF power amplifiers (RF PAs) and radio units or RRUs/RRHs containing RF chains, which convert digital signals from the BBU into RF signals for transmission over the antenna interface and vice versa. In addition, transmission power takes into account the power needed for low-noise amplification of receiving signals, filtering and frequency conversion, and operation of radio transceivers and duplexers used for separating uplink and downlink frequencies in uplink and downlink transmission.

Computation power refers to the power consumed by BBUs that perform digital signal processing for uplink and downlink transmission, signal modulation, coding and decoding, mobility management, resource allocation, and control/synchronization tasks for BSs, as well as communication between the CN and BS.

Additional power considers all remaining power consumption of the BS, including power losses during the conversion of electricity from the AC grid into DC, as well as power losses that occur during transfer between DC-DC converters and power needed for internal BS air cooling.

Also, the total EC of a BS or BS site consists of a fixed component that is independent of BS load and a variable EC component that directly depends on the level of BS load. Total BS or BS site power consumption is thus the sum of fixed and variable power consumption components. Fixed components of instantaneous BS power consumption are contributed by the power consumption of BS elements that consume energy even when the BS is not handling any traffic. Since active BSs need to transmit pilot signals to announce to UDs the presence of a network signal and maintain synchronization [15], BS elements that constantly consume energy are parts of the BBU and parts of the radio unit, DC-DC converters, and main power supply. Also, for BS sites that have cooling devices, the power consumption of cooling devices and power consumption of redundant power supply elements with other BS site elements used for the connection of the BS site with the CN, contribute to the fixed power consumption component of the BS site.

An increase in BS load that is directly impacted by the number of UDs simultaneously served by the BS contributes to the variable component of BS and BS site instantaneous power consumption. Generally, the BS and BS site instantaneous power consumption increases with the increased BS load. This means that the variable component of instantaneous BS power consumption will be higher for higher BS loads and vice versa. This is a consequence of the BS's need to engage a larger amount of resources in satisfying the transmission of a larger traffic load. These resources include BS transmission at the higher Tx power levels or engaging more cells, frequency resource blocks, data channels, power amplifiers, transceivers, RRU/RRH elements, BBU(s) capacity, etc. Engagement of a larger amount of BS resources contributes to the increase of variable BS power consumption components. Analyses presented in [41] and [42] point to the mostly linear relationship of variable and consequently overall BS or BS site instantaneous power consumption on BS load and Tx power, respectively.

3.7 The Impact of SDN Technology on Improving Network Energy Efficiency

Further advancement in the development of next-generation mobile networks will significantly depend on the implementation of the software-defined networking (SDN) concept. Implementation of SDN enables the creation of innovative and flexible mobile network architectures through the separation of data, control, and application layers of the network. Implementation of SDN thus enables the network to become directly programmable by separating the traffic management and decision-making tasks of the control plane from the data forwarding tasks of the data plane. Figure 3.5 illustrates the conceptual SDN architecture in mobile networks. The application plane (layer) (Figure 3.5), allows the implementation of network policies and applications that manage the network as a single logical entity.

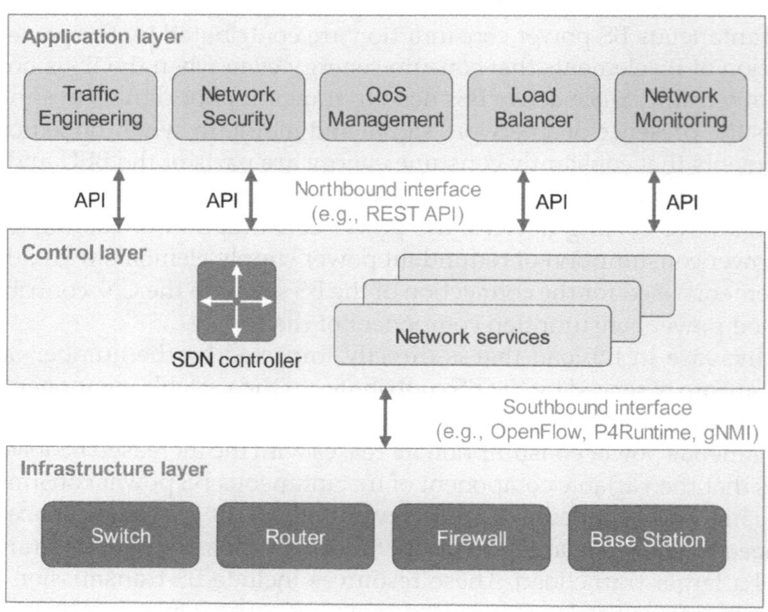

FIGURE 3.5
Conceptual presentation of SDN-based mobile network architecture with centralized management [43]

In the SDN architecture, network control is centralized within software-based SDN controllers operated by the SDN control software (Figure 3.5), which provides a complete view of the network. The network is presented to applications and policy management systems as a unified entity, allowing MNOs to manage the entire network from a single central point, independently of the networking equipment manufacturer. This approach significantly simplifies mobile network design and operation and reduces the complexity of network devices, which no longer need to process complex and versatile network protocols, and instead, network devices need only to follow operating instructions received from the SDN controller(s) [43]. The SDN concept thus not only increases mobile network flexibility and responsiveness to traffic variations but also contributes to the optimization of network EE. This is a consequence of reducing processing complexity in network elements which consequently contributes to the reduction in network element EC. Thus, the introduction of the SDN concept is particularly important in the context of the development of 5G and B5G networks, which are characterized with the need for the deployment of a large number of new network elements that must serve a constantly increasing number of connected UDs.

Additionally, the concept of network function virtualization (NFV) is commonly used in combination with SDN technology. It represents the transformative concept that enables in mobile networks the decoupling of network functions from dedicated hardware and implementing them as software running on general-purpose servers, mostly allocated in some cloud platforms. The NFV concept thus allows mobile network functions such as network traffic routing and switching to be performed on general-purpose server platforms rather than on specialized hardware. Dynamic scaling of mobile network resources according to space and time traffic variations, which is seen as one of the most important concepts for improving network EE, can be facilitated through the realization of mobile networks by a combination of SDN and NFV technologies [44]. This integration will be crucial for the development of future 6G networks, where the infrastructure is confronted with new challenges as a result of the increased performance requirements and the necessity for reduced EC.

Using a combination of NFV and SDN technologies, the authors in [45] developed an optimization model for energy allocation in 5G networks. Their method guarantees a high network QoS, while simultaneously enhancing network EE. The analysis of their optimization model demonstrated that sophisticated algorithms exploiting SDN and NFV concepts are capable of achieving substantial improvements in network EE, which is crucial for ensuring the long-term sustainability of 5G networks. Furthermore, optimization of traffic routing and placement of network functions can result in significant energy savings of 5G networks. For example, the authors in [46] predict that network EC can be reduced by 70% in certain operating scenarios, making this technology very efficient and the SDN/NFV concept the most important candidate for massive implementation in future mobile networks.

Additionally, research conducted in [47] concentrates on the integration of SDN technology into heterogeneous 5G networks, with a particular emphasis on enabling intelligent network resource management and ensuring network security. Their model for optimizing network performance through SDN demonstrates a substantial decrease in EC per network node, which proves the possibility of the sustainable development of future 5G systems. In addition, the study in [48] explores hybrid architectures that integrate SDN and software-defined radio (SDR) as a technology that uses software to control RF hardware within an NFV framework. The research findings indicate that flexibility offered by such network concepts in allocating central processing unit (CPU) resources and optimizing processes within cloud-based systems can result in an additional 20% network energy savings in comparison with conventional network management methods. This serves as an additional confirmation of the significance of implementing innovative technological solutions based on SDN and NFV in the development of energy-efficient and sustainable mobile networks of the future.

Overall, the integration of SDN technology into mobile networks can bring a significant advance in the realization of energy-efficient 5G and B5G networks. The SDN concept can achieve substantial energy savings, while simultaneously preserving high levels of performance and QoS through the implementation of sophisticated algorithms for optimizing network operations, dynamic mobile network resource allocation, and centralized management. The integration of SDN with other technologies, including, for example, NFV and machine learning, will create new opportunities for further network EC optimization, thereby reducing the energy footprint of mobile networks and ensuring their sustainable development in the future.

3.8 Network Slicing Technology as a Solution to Improving Network Energy Efficiency

Network slicing (NS) enables the creation of logical network segments or network slices tailored to specific user or application requirements. Such tailoring of network slice resources to the needs of specific network use cases can enable optimized use of network resources, which consequently can result in the improvement of mobile network EE. As presented in Figure 3.6, NS is becoming a critical technology in the realization of 5G and future 6G networks, due to the possibility of facilitating different ranges of use cases in mobile networks, such as the implementation of smart healthcare, smart

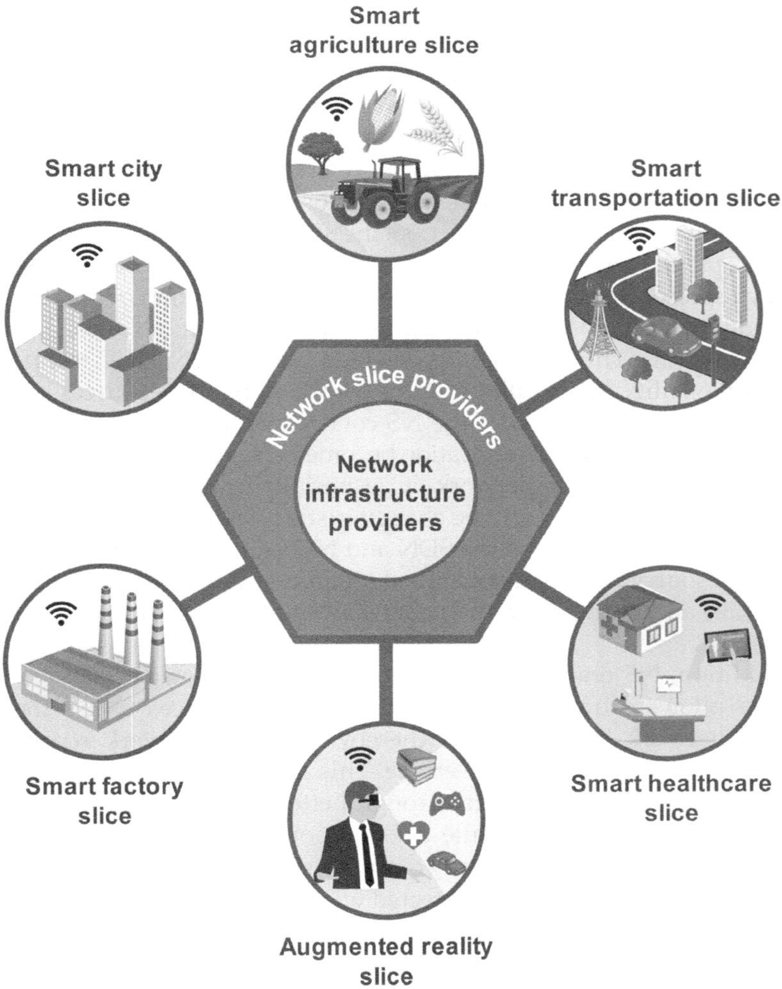

FIGURE 3.6
Possible use cases of network slicing in mobile networks [50]

agriculture, smart grid systems, smart transportation, and augmented reality. NS allows MNOs to share physical network infrastructure between different logical network segments, each of which can be tailored to specific NS customer requirements. Thus, NS can enable customized mobile network functionality to meet the specific needs of different use case applications and optimally exploit mobile network resources. Such NS capability enables the optimization of overall network resource usage, which consequently reduces excessive EC and improves the overall EE of the mobile network [49].

In addition, network segmentation through NS facilitates the implementation of the Network-as-a-Service (NaaS) concept, which, compared to the concept of realizing traditional non-sliced mobile networks, enables overall mobile networks or partial segments of mobile networks to be offered to the customers as a service [50]. This enables the creation of mobile virtual network operators that lease access to network resources and spectrum from traditional MNOs and provide services to customers under their own brand.

As presented in the previous section, the SDN/NFV technologies are essential concepts that enable the centralized management of network resources, thereby representing crucial technologies in the realization of NSs that need to be adaptively tailored to the requirements of specific NS use cases and applications [51]. This adaptability enabled by the SDN/NFV technologies ensures flexibility in the allocation of mobile network resources, which is essential for the realization of the NS concept. As mobile networks evolve, dynamic network resource allocation becomes essential to optimize network performance and reduce network EC, especially in high-traffic or high UD density conditions which will be characteristic of many future NS use cases. Thus, the implementation of the SDN and NFV concepts can enable efficient network resource management and adaptation to versatile conditions of different NSs, and at the same time contribute to the improvement of mobile network EE.

In addition, NS realized through the implementation of SDN and NFV concepts allows different industry verticals and services to use the same physical network infrastructure, thereby adapting network resources to meet the unique needs of each service. This approach guarantees reliability and maximizes energy savings through better and more rational exploitation of network resources, while also supporting low-latency applications and high-traffic services without compromising QoS. These characteristics are essential for delivering the expected wide range of versatile services over 5G and future 6G networks [52].

Overall, the implementation of the NS concept in mobile networks has the potential to be a crucial advancement in improving EE in the 5G and upcoming 6G networks. By creating virtual network segments designed for specific applications or services, NS optimizes network resource allocation, thereby contributing to reducing network EC for different network usage scenarios (such as those shown in Figure 3.6). In addition, real-time dynamic resource management, enabled by technologies such as SDN/NFV, makes it possible to maintain high network EE despite the varying QoS demands of specific NS. Hence, as 5G networks advance and 6G networks emerge, NS will become an important concept in meeting the targeted performance and sustainability goals of mobile networks.

References

1. GSMA, "Mobile net zero state of the industry on climate action 2021," 2021. Accessed: Aug. 17, 2024. [Online]. Available: https://www.gsma.com/solutions-and-impact/connectivity-for-good/external-affairs/wp-content/uploads/2021/04/Mobile-Net-Zero-State-of-the-Industry-on-Climate-Action.pdf
2. C. Freitag, M. Berners-Lee, K. Widdicks, B. Knowles, G. S. Blair, and A. Friday, "The real climate and transformative impact of ICT: A critique of estimates, trends, and regulations," 2021. doi: 10.1016/j.patter.2021.100340.
3. Global e-Sustainability Initiative (GeSI), "SMARTer2030-ICT solutions for 21st century challenges," *Global e-Sustainability Initiative (GeSI)*, Brussels, Belgium, 2015.
4. Ltd. Huawei Technologies Co., "Green 5G white paper," 2020, Accessed: Aug. 18, 2024. [Online]. Available: https://www-file.huawei.com/-/media/corp2020/pdf/tech-insights/1/green_5g_white_paper_en_v2.pdf
5. J. Lorincz, L. Chiaraviglio, and F. Cuomo, "A measurement study of short-time cell outages in mobile cellular networks," *Computer and Communications*, vol. 79, 2016. doi: 10.1016/j.comcom.2015.12.006.
6. M. H. Alsharif, J. Kim, and J. H. Kim, "Green and sustainable cellular base stations: An overview and future research directions," 2017. doi: 10.3390/en10050587.
7. W. Hao, M. Zeng, Z. Chu, S. Yang, and G. Sun, "Energy-efficient resource allocation for mmwave massive MIMO HetNets with wireless backhaul," *IEEE Access*, vol. 6, 2017. doi: 10.1109/ACCESS.2017.2783544.
8. 5G Americas, "Energy efficiency and sustainability in mobile communications networks - white paper," Dec. 2023. Accessed: Nov. 04, 2024. [Online]. Available: https://www.5gamericas.org/wp-content/uploads/2023/12/Energy-Efficiency-and-Sustainability-in-Mobile-Communications-Networks-WP.pdf
9. European Telecommunications Standards Institute (ETSI), *ES 203 228 V1.3.1 (2020-10): Environmental Engineering (EE); Assessment of mobile network energy efficiency*, Sophia Antipolis, France: ETSI, Oct. 2020.
10. 3GPP, "TR 32.972 (V16.1.0) (2019-09) Study on system and functional aspects of energy efficiency in 5G networks (Release 16)," 2019. Accessed: Aug. 24, 2024. [Online]. Available: https://portal.3gpp.org/desktopmodules/Specifications/SpecificationDetails.aspx?specificationId=3279
11. ITU-T, "L.1331 (01/2022) Assessment of mobile network energy efficiency," 2022. Accessed: Aug. 24, 2024. [Online]. Available: https://www.itu.int/rec/T-REC-L.1331-202201-I/en
12. ETSI, "TS 128 310 (V18.4.0) LTE; 5G; Management and orchestration; Energy efficiency of 5G," 2024. Accessed: Aug. 18, 2024. [Online]. Available: https://www.etsi.org/deliver/etsi_ts/128300_128399/128310/18.04.00_60/ts_128310v180400p.pdf
13. NGMN Alliance, "Green future networks - network energy efficiency," 2021. Accessed: Aug. 22, 2024. [Online]. Available: https://www.ngmn.org/wp-content/uploads/211009-GFN-Network-Energy-Efficiency-1.0.pdf

14. J. Lorincz, Z. Klarin, and D. Begusic, "Modeling and analysis of data and coverage energy efficiency for different demographic areas in 5G networks," *IEEE Systems Journal*, vol. 16, no. 1, 2021. doi: 10.1109/JSYST.2021.3074100.
15. P. Kaur, R. Garg, and V. Kukreja, "Energy-efficiency schemes for base stations in 5G heterogeneous networks: A systematic literature review," 2023. doi: 10.1007/s11235-023-01037-x.
16. M. Heskamp, R. Schiphorst, and K. Slump, "Public safety and cognitive radio," in *Cognitive Radio Communications and Networks*, 2010. doi: 10.1016/b978-0-12-374715-0.00016-2.
17. M. I. Salman, C. K. Ng, and N. K. Noordin, "Energy- and spectral-efficient wireless cellular networks," in *Lecture Notes of the Institute for Computer Sciences, Social-Informatics and Telecommunications Engineering*, 2012. doi: 10.1007/978-3-642-33368-2_15.
18. Q. D. Vo, J. P. Choi, H. M. Chang, and W. C. Lee, "Green perspective cognitive radio-based M2M communications for smart meters," in *2010 International Conference on Information and Communication Technology Convergence, ICTC 2010*, 2010. doi: 10.1109/ICTC.2010.5674821.
19. X. Zhang, H. Qi, X. Zhang, and L. Han, "Spectral efficiency improvement and power control optimization of massive MIMO networks," *IEEE Access*, vol. 9, 2021. doi: 10.1109/ACCESS.2021.3051170.
20. M. P. Mota, D. C. Araujo, F. H. Costa Neto, A. L. F. De Almeida, and F. R. Cavalcanti, "Adaptive modulation and coding based on reinforcement learning for 5G networks," in *2019 IEEE Globecom Workshops, GC Wkshps 2019 - Proceedings*, 2019. doi: 10.1109/GCWkshps45667.2019.9024384.
21. J. Lorincz, I. Ramljak, and D. Begušić, "Analysis of the impact of detection threshold adjustments and noise uncertainty on energy detection performance in MIMO-OFDM cognitive radio systems," *Sensors*, vol. 22, no. 2, 2022. doi: 10.3390/s22020631.
22. J. Lorincz, I. Ramljak, and D. Begusic, "Algorithm for evaluating energy detection spectrum sensing performance of cognitive radio MIMO-OFDM systems," *Sensors*, vol. 21, no. 20, 2021. doi: 10.3390/s21206881.
23. J. Lorincz, I. Ramljak, and D. Begušić, "A survey on the energy detection of ofdm signals with dynamic threshold adaptation: Open issues and future challenges," 2021. doi: 10.3390/s21093080.
24. H. Holma, A. Toskala, and T. Nakamura, *5G Technology: 3GPP New Radio*. John Wiley & Sons, 2019. doi: 10.1002/9781119236306.
25. Y. Zhang and A. Årvidsson, "Understanding the characteristics of cellular data traffic," in *CellNet'12 - Proceedings of the ACM Workshop on Cellular Networks: Operations, Challenges, and Future Design*, 2012. doi: 10.1145/2342468.2342472.
26. S. Kemp, "Digital 2023 July global statshot report." Accessed: Aug. 18, 2024. [Online]. Available: https://datareportal.com/reports/digital-2023-july-global-statshot
27. Ericsson, *Exploring How Traffic Patterns Drive Network Evolution: Extract from the Ericsson Mobility Report*. Stockholm, Sweden: Telefonaktiebolaget LM Ericsson, Jun. 2023.
28. Q. Ou *et al.*, "A method of dynamic resource adjustment for 5G network slice," in *Advances in Intelligent Systems and Computing*, 2021. doi: 10.1007/978-981-15-3753-0_91.

29. G. Wu and D. Zeng, "Optimization of network resource management based on software-defined networking in the 5G environment," *Journal of Supercomputing*, vol. 78, no. 15, 2022. doi: 10.1007/s11227-022-04547-8.
30. S. E. Elayoubi, L. Saker, and T. Chahed, "Optimal control for base station sleep mode in energy efficient radio access networks," in *Proceedings - IEEE INFOCOM*, 2011. doi: 10.1109/INFCOM.2011.5934881.
31. K. Dufková, M. Bjelica, B. Moon, L. Kencl, and J. Y. Le Boudec, "Energy savings for cellular network with evaluation of impact on data traffic performance," in *2010 European Wireless Conference, EW 2010*, 2010. doi: 10.1109/EW.2010.5483431.
32. X. Guo, Z. Niu, S. Zhou, and P. R. Kumar, "Delay-constrained energy-optimal base station sleeping control," *IEEE Journal on Selected Areas in Communications*, vol. 34, no. 5, 2016. doi: 10.1109/JSAC.2016.2520221.
33. S. R. Samal, "Interference management techniques in small cells overlaid heterogeneous cellular networks," 2018. doi: 10.13052/jmm1550-4646.1432.
34. N. Trabelsi, L. Chaari Fourati, and C. S. Chen, "Interference management in 5G and beyond networks: A comprehensive survey," 2024. doi: 10.1016/j.comnet.2023.110159.
35. F. Baccelli, M. Klein, M. Lebourges, and S. Zuyev, "Stochastic geometry and architecture of communication networks," *Telecommun Syst*, vol. 7, no. 1–3, 1997. doi: 10.1023/a:1019172312328.
36. J. Lorincz, Z. Klarin, and D. Begusic, "Advances in improving energy efficiency of fiber–wireless access networks: A comprehensive overview," 2023. doi: 10.3390/s23042239.
37. Huawei, "Green 5G: Building a sustainable world," 2020. Accessed: Aug. 18, 2024. [Online]. Available: https://www.huawei.com/en/public-policy/green-5g-building-a-sustainable-world
38. J. Lorincz, A. Capone, and J. Wu, "Greener, energy-efficient and sustainable networks: State-of-the-art and new trends," 2019. doi: 10.3390/s19224864.
39. M. Usama and M. Erol-Kantarci, "A survey on recent trends and open issues in energy efficiency of 5G," 2019. doi: 10.3390/s19143126.
40. X. Ge, J. Yang, H. Gharavi, and Y. Sun, "Energy efficiency challenges of 5G small cell networks," *IEEE Communications Magazine*, vol. 55, no. 5, 2017. doi: 10.1109/MCOM.2017.1600788.
41. J. Lorincz, T. Garma, and G. Petrovic, "Measurements and modelling of base station power consumption under real traffic loads," *Sensors*, vol. 12, no. 4, 2012. doi: 10.3390/s120404281.
42. G. Micallef, P. Mogensen, and H. O. Scheck, "Cell size breathing and possibilities to introduce cell sleep mode," in *2010 European Wireless Conference, EW 2010*, 2010. doi: 10.1109/EW.2010.5483401.
43. Open Networking Foundation, "Software-defined networking: The new norm for networks [white paper]," *ONF White Paper*, 2012.
44. S. Iyer, "Virtualisation of network functions and the SDN: : Improving the economics of the network.," *Journal of Telecommunications and the Digital Economy*, vol. 2, no. 2, 2014. doi: 10.18080/jtde.v2n2.332.
45. S. Yadav and S. Nanivadekar, "Hybrid optimization assisted green power allocation model for QoS-driven energy-efficiency in 5G networks," *Cybernetics and Systems*, 2023. doi: 10.1080/01969722.2023.2175147.

46. R. Moosavi, S. Parsaeefard, M. A. Maddah-Ali, V. Shah-Mansouri, B. H. Khalaj, and M. Bennis, "Energy efficiency through joint routing and function placement in different modes of SDN/NFV networks," *Computer Networks*, vol. 200, 2021. doi: 10.1016/j.comnet.2021.108492.
47. O. M. Ebadati E and H. R. Ebadati E, "Implementing software-defined networks in heterogeneous 5G communications to provide security and intelligent resource management," in *Lecture Notes on Data Engineering and Communications Technologies*, vol. 124, 2022. doi: 10.1007/978-3-030-97610-1_54.
48. B. Chen, F. Slyne, and M. Ruffini, "Energy efficient SDN and SDR joint adaptation of CPU utilization based on experimental data analytics," in *IEEE International Conference on Communications*, 2023. doi: 10.1109/ICC45041.2023.10279620.
49. J. Lorincz, A. Kukuruzović, and Z. Blažević, "A comprehensive overview of network slicing for improving the energy efficiency of fifth-generation networks," May 01, 2024, *Multidisciplinary Digital Publishing Institute (MDPI)*. doi: 10.3390/s24103242.
50. L. U. Khan, I. Yaqoob, N. H. Tran, Z. Han, and C. S. Hong, "Network slicing: Recent advances, taxonomy, requirements, and open research challenges," *IEEE Access*, vol. 8, 2020. doi: 10.1109/ACCESS.2020.2975072.
51. M. Condoluci, F. Sardis, and T. Mahmoodi, "Softwarization and virtualization in 5G networks for smart cities," in *Lecture Notes of the Institute for Computer Sciences, Social-Informatics and Telecommunications Engineering, LNICST*, 2016. doi: 10.1007/978-3-319-47063-4_16.
52. X. Foukas, G. Patounas, A. Elmokashfi, and M. K. Marina, "Network slicing in 5G: Survey and challenges," 2017. doi: 10.1109/MCOM.2017.1600951.

4

Optimization of Network Energy Consumption

Optimization of energy consumption (EC) in wireless communication networks represents one of the key challenges of the modern telecommunications industry. Given the increasing demand for wireless services and the accelerated increase in the number of connected devices, energy efficiency (EE) is becoming a critical factor for the sustainability and economy of mobile network operations [1]. According to estimations in [2, 3], the radio access network (RAN) accounts for the largest portion of total energy consumption in mobile networks. As a result, the RAN has become the primary focus of research efforts in both industry and academia, with the aim of reducing energy usage and enhancing the EE of mobile networks.

Improving RAN energy efficiency can be achieved through various approaches, including hardware optimization, implementation of energy-efficient algorithms and protocols, and use of advanced RAN resource management techniques. Various strategies such as dynamic RAN resource management, base station (BS) transmit (Tx) power optimization, and implementation of small BS cells have already shown significant potential in reducing RAN energy consumption.

To further enhance EE in mobile networks, additional energy-saving strategies have been developed. According to [4], these strategies include improvements at the level of hardware components, the application of sleep mode techniques, BS Tx power optimization, different approaches to deploying new network elements (such as different BS types), and the use of renewable energy sources (Table 4.1). Each of these strategies has its specific advantages and limitations. The following subsections provide a comparison of these approaches (Table 4.1), with examples of their application, as well as the key advantages and challenges in implementation. A comprehensive review of techniques and strategies for improving EE of next-generation mobile communication networks, that includes an analysis of key design elements with an emphasis on BSs as the largest energy consumer in RAN [5], is presented in subsequent subsections.

DOI: 10.1201/9781003615866-4

TABLE 4.1

Comparison of Energy-Efficient Strategies for Operation of Mobile Networks [4]

Energy-saving Strategy	Example	Advantages	Limitations
Improvement of individual components	Enhanced power amplifiers, BBUs, and other BSS hardware	Significant savings, straightforward implementation	Incremental improvements, high hardware replacement costs
Sleep (power saving) techniques	Selective on and off sleep mode switching of complete BSs or BS parts	Substantial energy savings, cost-effective to test and deploy	Balancing network performance and energy savings can be challenging with current network resources lacking accuracy for optimal scheduling
Radio transmission and resource optimization	Cognitive radio, cooperative transmission, MIMO and mMIMO transmission	Cost-effective approach, supports diverse use cases	Performance and EE trade-offs, errors caused by inaccuracy of spectrum sensing due to dynamic wireless channel conditions
Network deployment	Augmentation of macro BSs with micro, pico, and femto BSs	Low setup costs, user-focused designs, significant potential for energy savings	Interference and RAN resource coordination complexities
Power supply with renewable energy sources	Solar, wind, and hydro energy for powering BS sites	Sustainable long-term solution for off-grid BSs	High installation costs and limited feasibility for urban grid-connected BS sites

4.1 Energy Efficiency Optimization of Base Station Components

Given the increasing demands for higher network capacity and faster data transfer speeds, enhancing the EE of the individual BS or base station site (BSS) hardware components becomes crucial for minimizing the overall energy consumption of mobile networks. Figure 4.1 presents the EC distribution of a typical macro BSS. The largest share of EC in BSs is attributed to radio-frequency power amplifiers (RF PAs), which amplify signals to the required levels for transmission, and to feeder links, which subsequently deliver these amplified signals to the antennas. In contrast, baseband units (BBUs) are responsible for processing and digitizing signals before they are transmitted at antenna interfaces, which is essential for efficient communication between BSs and user devices (UDs). Despite the fact that their responsibilities are different, both components (PAs and BBUs) contribute significantly to the overall EC of the BSS and consequently are seen as essential components for EC optimization.

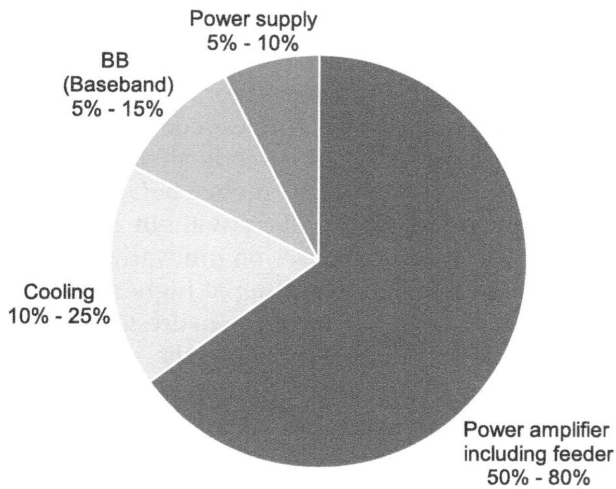

FIGURE 4.1
The energy consumption distribution of macro base station site components [6]

Beyond hardware-level optimization, the EE of BSs and BSSs can be further enhanced through advanced software techniques and algorithms that enable intelligent network resource management. In addition, the use of machine learning (ML) and other artificial intelligence (AI) methods can facilitate real-time adjustment of operating parameters, further increasing the efficiency of energy consumption management in mobile networks.

The integration of advanced technologies and methods to optimize EE at the component level of individual BSs and BSSs contributes to reducing mobile network EC, while enabling mobile network operators to meet the growing sustainability demands of next-generation networks, ensuring network long-term functionality and superior quality of service (QoS).

4.1.1 Radio Frequency Power Amplifiers

In the context of mobile network EE optimization, RF PAs are important components of BSs, since their power consumption significantly exceeds the power consumption of other BS components (Figure 4.1). The basic function of an RF PA is to amplify the radio signal to the power level required to transmit the signal to the receiver [7]. Thus, the overall EE of the RF transmitter of the BS is directly influenced by the performance of RF PAs, making this component indispensable for improving the EE of BSs in mobile networks.

Due to the growing demands for improving mobile networks' EE and reducing operating expenditure (OPEX), research in areas dedicated to improving PA EE is becoming increasingly important. Research in paper [8] examines in detail the development trends of RF PA, including the globalization of PA development, inter-institutional cooperation in the application of

PAs, and innovations in PA architecture. This combined approach to RF PA development contributes to faster progress and improvement of technology for RF PA development and its implementation in BSs of the fifth-generation (5G) and future sixth-generation (6G) wireless communication systems.

Since the performance of the RF PA directly affects the efficiency of each BS RF transmitter, the new BS in the 5G mobile network requires advanced PA architectures with improved efficiency without compromising the linearity of the PA. The study in [9] focuses on innovative RF PA architectures that address the challenges of transmitting at higher frequencies and with larger bandwidths, along with the need to ensure greater PA EE of BSs in 5G networks. These architectures emphasize the use of Doherty amplifiers, wireless signal envelope tracking (ET), and predistortion techniques to improve linearity and reduce PA power consumption. This evolution of RF PA architectures reflects the need to balance performance, cost, and EE in next-generation wireless communication systems.

In addition to traditional PA optimization techniques, advanced methods such as the use of AI in the optimization of PA EE are attracting increasing research interest. To optimize the EE of RF PAs in 5G BSs, the study in [10] presents an innovative learning approach that employs deep neural networks (DNNs) to compensate for amplifier nonlinearity and memory effects. The proposed learning framework in the study uses DNNs to simulate amplifier characteristics and generate digital predistortion (DPD) to mitigate the detrimental effects of spectral regrowth and nonlinearity. This approach resulted in significant improvements over traditional methods, including a 34% reduction in spectral regrowth in cases of severe distortion. The effectiveness of this approach makes it crucial for improving the PA EE of BSs in 5G networks.

In summary, the significant power demands of RF PAs (Figure 4.1) make them a critical component in enhancing EE within BSs. Advancements in RF PA design, including envelope tracking, predistortion, and Doherty amplifiers, are essential for enhancing PA efficiency without sacrificing performance, particularly as 5G networks and especially 6G networks are envisioned in the future to operate at high frequencies (in millimeter-wave bands) for ensuring high bandwidths. Furthermore, the utilization of intelligent optimization techniques has demonstrated substantial improvements in the reduction of nonlinearity and spectral regrowth through the implementation of innovative AI approaches, including DNNs. However, supporting the development of sustainable and energy-efficient next-generation mobile networks will necessitate further advancements in RF PA technology.

4.1.2 Baseband Units

Baseband units are essential components of a BS responsible for signal processing, protocol layer management, resource allocation, synchronization, and interfacing the BS with wireless transmission BS elements and core network (CN) (Figure 3.4). BBUs handle data processing before it is converted to radio frequencies and transmitted via antennas. Due to their complexity and constant operation, the EE of BBUs has become an important point in efforts to reduce the overall power consumption of BSs, especially in the context of next-generation mobile networks.

At the hardware level, field programmable gate array (FPGA) technology has a key role in optimizing the EE of BBUs that contain FPGA circuits. FPGA technology enables the acceleration of specific data and signal processing tasks with the flexibility to change the BBU configuration according to network data and signaling traffic needs. The use of FPGA technology facilitates dynamic BBU reconfiguration and parallel processing, which optimizes BBU energy consumption during different operating conditions and adapts the operation of BBUs to current network needs. This technology plays a pivotal role in significantly reducing EC within BBUs, a benefit that is especially critical in architectures like cloud radio access network (C-RAN) systems, where EE is paramount [11, 12]. However, further enhancements in the operating performance of FPGA chipsets are needed to support a diversity of upcoming network services, that will be enabled in advanced 5G networks and future 6G networks.

In addition, important methods for improving the EE of BBUs include dynamic voltage and frequency scaling (DVFS) and dynamic power management (DPM) of BBU electronic components [5, 13]. DVFS technology allows the voltage and frequency of the BBU processor(s) to be adjusted according to the current data processing needs, thus reducing power consumption when the demand for BBU processing resources is low. DPM technology, on the other hand, puts the BBU processor(s) into a sleep state in the periods lacking network traffic activity, thus further reducing processors and consequently BBU power consumption. The combination of these techniques optimizes BBU resources, especially in conditions of variable network traffic, which is crucial for reducing OPEX and the carbon footprint of equipment used in mobile networks.

Furthermore, at the BBU software level, BS EE can be additionally enhanced through the implementation of advanced resource management algorithms and predictive resource allocation, that is based on anticipated traffic variations that the BBU must handle [14, 15]. In this context, leveraging ML techniques offers the potential for real-time adjustments of operating parameters and predictive resource allocation by utilizing historical data and real-time network conditions [16]. By integrating these methods, it is possible to manage BBU operations more intelligently, which ultimately results in the optimization of BS energy consumption (EC) and the improvement of overall mobile network energy efficiency (EE).

4.2 Energy-Saving Sleep Mode Strategies

Sleep mode techniques in BSs are energy-saving strategies that dynamically adjust the power consumption of the BS or its individual components based on the current traffic load and demand for network resources. Thus, sleep mode strategies allow network components, such as entire BSs or parts of BSs, to enter a low-power consumption state when there is no need to operate at the full capacity of available resources [17]. These techniques aim to reduce EC during periods of low activity, while ensuring that QoS and user connectivity are not compromised. Such techniques have huge potential to enable substantial energy savings and are crucial for reducing the EC of mobile networks, especially during periods of low traffic in certain parts of the mobile network [18, 19]. Although starting to be implemented in 4G networks, the full and massive implementation of these techniques is envisioned in 5G and 6G networks. Given the increasing demands for EE in next-generation networks, the application of sleep mode techniques will be the standard operating option of modern BSs.

Generally, BS sleep mode techniques can be divided into a few key categories that include complete BS sleep modes and component-level BS sleep modes. In complete BS sleep mode strategies, entire macro or small cell BSs are put in sleep mode of operation, according to traffic variations in specific parts of the network. This technique is complemented by the so-called "cell zooming" concept where low-traffic demand areas "zoom-out" by powering down complete BSs or reducing BS Tx power, while BSs that remain active temporarily cover the "zoom-out" area by increasing RF Tx power in order to maintain network coverage and capacity. Component-level sleep modes are applied to individual components within a BS to save energy. Individual components that are mostly seen to be included in sleep mode techniques are RF chains with PAs, BBUs, RF channels with resource blocks (RBs), and remote radio units/heads (RRUs/RRHs) with corresponding antenna elements. Additionally, some sleep mode techniques can enable different depths of sleep states such as micro-sleep or deep-sleep modes of operation. In the micro-sleep mode of operation, components like PAs, radio channels, RBs, or parts of RRU/RRH elements are temporarily turned off for milliseconds between data transmissions, which can contribute to reducing overall energy use of BSs during operation without affecting network performance. In the deep sleep mode of BS component operation, during prolonged low-traffic periods, entire components or subsystems can enter a deeper sleep state, consuming only minimal energy to stay ready for reactivation. The depth of the BS sleep mode has an impact on the amount of energy saving, and although deeper sleep modes of BS components or complete BS can bring higher energy savings, the longer transition from deeper sleep to active mode of BSs imposes higher challenges on ensuring the overall network (BS) performance and QoS.

4.2.1 Overview of Sleep Mode Techniques

Different techniques (summarized in Table 4.2) can be applied to practically implement sleep mode methods in BSs. The authors in [4] highlight several key approaches that include various methods of user association with the network, self-organizing networks (SONs), dynamic BS Tx power adjustments, advanced traffic prediction, the implementation of heterogeneous networks, and BS cooperative communications. The efficient implementation of these methods depends on the interaction between BSs in the network, which enables the redistribution of resources from idle BSs or corresponding components to surrounding active BSs.

Implementation of these methods mainly depends on the BS controller to guarantee the transfer of radio channel resources while preserving the required network QoS. Also, for each application of EC optimization using sleep mode techniques, it is important to monitor the signal-to-noise ratio (SNR) on both user and network devices, as putting BSs or BS components into sleep mode may affect user experience, especially in high-load scenarios [28]. Maintaining optimal SNR is essential because it directly affects the quality and reliability of network connections. A high SNR ensures a robust,

TABLE 4.2

Overview of Mobile Network Energy-Saving Techniques Based on Sleep Mode Methods

Reference	Sleep Mode Methods	Description
[20]	User association, resource allocation	Optimization of user association with BSs to reduce EC and maintain service quality during sleep mode
[21, 22]	SON concept, automated network management, BS configuration adjustment	Automated management of network resources without human intervention, supporting sleep mode and optimizing energy efficiency
[23]	Transmit power adjustment, cell size adaptation	Adjusting cell size according to traffic conditions, modifying transmit power to reduce EC and improve network EE
[24]	Traffic prediction, sleep state adaptation	Using historical data to predict future traffic patterns and optimize BS sleep states
[25–27]	EE optimization through BS cooperation among various types of BSs in HetNets	Implementing sleep mode techniques in HetNets consisting of different types of BSs to optimize EE and maintain high service quality
[28, 29]	Cooperative communications based on coordinated multi-point (CoMP) transmission, power management, BS deactivation, service quality maintenance	Applying CoMP transmission strategies between BSs to optimize EE. Selectively deactivating BSs during low-traffic periods and redirecting traffic to remaining active BSs saves energy without compromising QoS

distinct signal in relation to background noise, enabling reliable data transfer while improving the overall user experience. On the other hand, a low SNR can lead to connection drops, reduced data rates, and decreased QoS, particularly during periods of increased network demand or BS component reactivation from sleep mode of operation. Consequently, by carefully balancing SNR with the sleep mode strategies, mobile network operators can attain energy savings without sacrificing QoS.

In the current generation of mobile networks, sleep mode techniques are essential to optimize EC by selectively deactivating entire BSs or BS components during periods of low traffic. These methods can provide substantial energy savings, particularly as expectations related to network performance will continue to grow for next-generation network architectures, such as ultra-dense heterogeneous network (HetNet) architectures. However, the success of sleep mode methods depends on a variety of strategies, including UD association, implementation of the self-orginizing network (SON) concept, and cell size adjustments, all of which enable networks to dynamically adapt to fluctuating traffic loads while preserving QoS. Furthermore, effective sleep management is further facilitated by advanced techniques, including algorithm-driven UD and resource reallocation, collaboration strategies between macro and micro BSs, and traffic prediction. These sleep mode methods should enable mobile network operators to achieve significant energy savings while maintaining a seamless user experience, making sleep mode methods crucial for building sustainable, flexible, and energy-efficient mobile networks of the future. The following subsections present each method.

4.2.2 Sleep Mode Methods Based on Optimized User Association

When components or entire BSs enter the idle or energy-saving sleep state, UDs connected to these BS(s) are redirected to remaining active BS(s) that maintain uninterrupted signal coverage and prevent QoS degradation. Figure 4.2 presents micro BS sleep mode-based optimization of user device association. Optimizing UD association with BS(s), which is determined by UD locations, signal quality, and traffic load, is a crucial strategy for reducing network EC during the sleep mode of BS(s) or their components [20]. This strategy includes dynamic user reallocation and load balancing across active BSs, coordination between macro and small cells, and performance tradeoff management (Table 4.2). Optimized dynamic UD reallocation and load balancing across active BSs ensures that users are redistributed efficiently, reducing the need for keeping redundant BSs active. This reduces the overall network EC by concentrating traffic on fewer active BSs or BS components. When applying the sleep mode strategy in HetNets, coordination between macro and small cells is mostly based on macrocells handling the load of small cells that change the operating state into sleep mode.

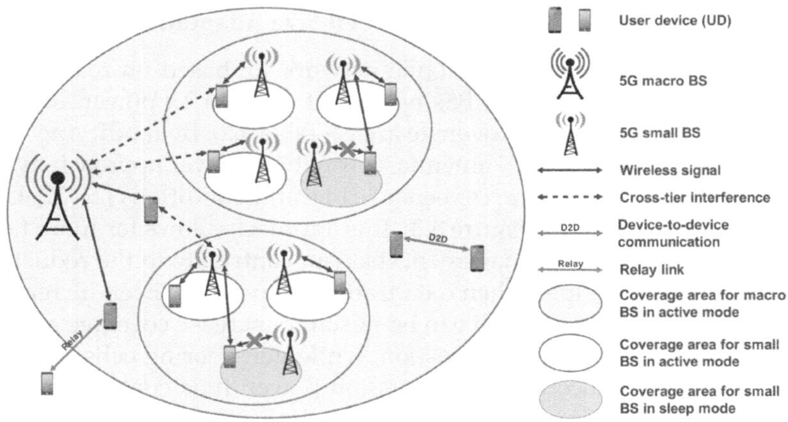

FIGURE 4.2
BS sleep mode-based optimization of user device association

Optimized user association ensures that small cells are used only when needed, reducing energy use during off-peak hours. Performance trade-off management is important when implementing the sleep mode strategies since keeping the balance between sleep mode activation/deactivation and BS utilization is based on setting thresholds for acceptable QoS parameters (e.g., throughput, delay), which need to ensure that energy savings do not come at the expense of user experience.

4.2.3 Sleep Mode Methods Based on the SON Concept

The SON concept, outlined in the technical standard [30], facilitates automated network management and BS configuration modifications without human involvement. SONs leverage intelligent algorithms and real-time data to manage the sleep states of core and radio parts of mobile network resources more efficiently (Table 4.2). This concept ultimately improves the functionality of the sleep mode technique by optimizing the BS activity state, transmission rate, or Tx power based on a prediction of the UD location and corresponding demands for network capacity [21, 22]. When new BSs are deployed in the mobile network, the main goal of the SON approach is to automate their configuration and optimize their operating parameters (e.g., Tx power, frequency allocation, RB allocation, number of PAs, BBUs) to ensure energy-efficient operation from the start. In the event of a deactivating BS or BS component, SONs can reconfigure the network to reroute traffic through the most energy-efficient paths. Thus, SONs use real-time data to optimize cell coverage and capacity planning, ensuring that the network is neither over-provisioned nor under-provisioned (Figure 4.2). This enables reducing unnecessary EC while maintaining optimal network performance.

4.2.4 Sleep Mode Methods Based on Cell Size Adaptation

Cell size adaptation enhances mobile network EE based on real-time traffic demands by ensuring that BSs operate at optimal Tx power and capacity levels and have optimal coverage areas (Table 4.2). By modifying the tilt, height, or Tx power of the BS antenna, this network layer approach dynamically modifies the BS cell size in response to traffic conditions [23, 31]. Unlike switching BSs on and off (Figure 4.3), this approach allows for more flexible and refined traffic load management that can contribute to the reduction of network EC. For example, when data traffic in a particular cell increases, BS transmission power in that cell can be raised to increase coverage and especially capacity and prevent congestion, while neighboring cells with lower traffic loads can reduce their transmission power, thus contributing to the network's energy savings (Figure 4.2). By such shrinking of BS cell coverage areas, a network can offload users to neighboring cells, allowing underutilized BSs to enter sleep mode or turn off certain components (e.g., RF chains, RBs, or PAs).

Additionally, cell zooming allows for adaptive BS cell sectorization in which, during low-traffic periods, fewer BS sectors can be active and thus reduce the EC of the BS, while in the peak traffic periods, additional sectors can be activated to maintain QoS. Such dynamic adjusting of cell sizes through cell activity states contributes to the decrease in interference between neighboring cells, which consequently reduces BS EC due to less energy needed for signal processing and retransmissions. Also, cell size adaptation through focusing the energy of wireless signals on active users

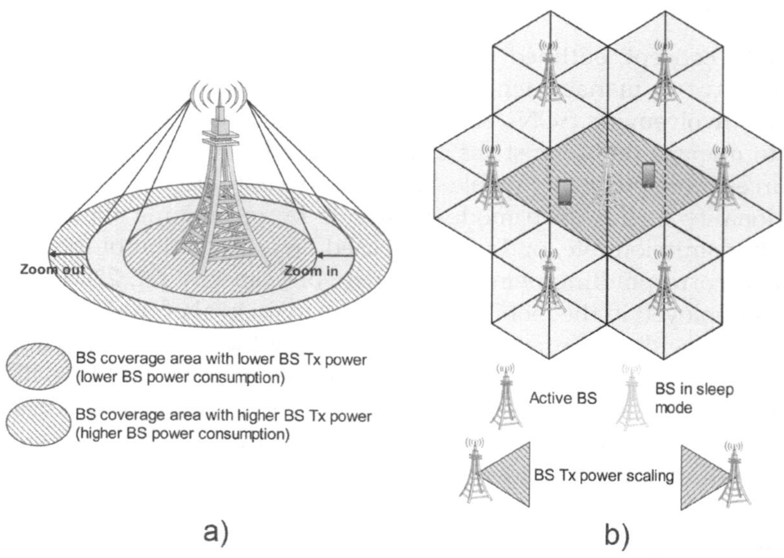

FIGURE 4.3
BS transmit power optimization through (a) the concept of single BS cell-zooming, (b) the concept of multiple BSs cell-zooming for covering uncovered areas of BS in sleep mode

(known as beamforming) allows BSs to prioritize coverage in areas where users are concentrated, which enables reducing energy waste in empty or sparsely populated areas. Thus, cell size adaptation techniques will be in practice a broadly accepted approach among BS sleep mode techniques [4].

4.2.5 Sleep Mode Methods Based on Traffic Prediction

Traffic prediction has an important role in improving mobile network EE and reducing EC in the context of implementing BS sleep mode techniques. By analyzing historical traffic data, network operators can forecast future traffic patterns. By accurately forecasting traffic patterns and user demands, network operators can make informed decisions about when and where to activate sleep modes for entire BSs or BS components (Table 4.2).

Traffic prediction techniques are based on the development of algorithms able to analyze historical and real-time data to forecast periods of low traffic in specific network areas. During these periods, BSs or their components, such as BBUs, RRHs/RRUs, power amplifiers, RBs, or RF chains, can be transitioned to sleep mode (Figure 4.2). Based on historical seasonal traffic patterns (e.g., holidays), hourly (e.g., rush hour vs. early morning), and daily traffic variations (working vs. weekend day), traffic prediction enables dynamic scheduling of when BSs or specific BS components should enter sleep mode and when they should wake up to handle rising traffic. This ensures that BSs are active only when needed and thus accurate traffic forecasts allow the network to allocate only the necessary resources, avoiding the activation of additional BSs or BS components during periods when they are not required. Also, accurate traffic prediction allows the network to prepare for handoffs by forecasting user mobility and ensuring that active BSs in adjacent areas are ready to handle additional traffic (Figure 4.2). This enables the network to anticipate when BSs need to wake up, allowing for smooth transitions between sleep and active states, while minimizing energy spikes associated with abrupt BS wake-ups.

Besides the stated advantages that traffic prediction techniques can bring to network resource management, the traffic prediction approaches must assist in the preservation of network QoS and the adaptation to varying network conditions. Therefore, networks that can make precise traffic predictions can optimize EC and enhance overall resource management by employing methods that integrate information about spatial and temporal data variations, which can result in substantial energy savings [24].

4.2.6 Sleep Mode Methods Based on Heterogeneous Networks

In ultra-dense heterogeneous mobile networks composed of a large number of various types of BSs (such as macro, micro, pico, femto), the implementation of sleep mode techniques can be particularly effective [32] (Figure 4.2). Generally, HetNets are more energy efficient than traditional (up to 4G) networks relying solely on macro BSs due to the lower power consumption of small (micro, pico, femto) BSs [25]. However, while small BSs (micro, pico, and femto) have lower overall EC compared to macro BSs, their number will be large in current 5G and especially next-generation (6G) HetNets (and even extremely high in some urban dense areas), and this generally contributes to a significant increase in the EC of the network [26]. As a consequence, the implementation of sleep mode techniques in HetNets is essential, as it enables the dynamic deactivation of BSs or BS parts during periods of low traffic (Figure 4.2). This dynamic scheduling of BSs or BS components can result in substantial energy savings despite the large number of versatile BSs installed in the network [17, 19].

HetNets support the implementation of all previously presented sleep state techniques that are related to intelligent UD association, SON concept, cell size adaptation, and traffic prediction. Researchers have examined sleep mode strategies that prioritize the preservation of QoS in order to optimize EC in HetNets. A method that emphasizes the configuration of BS sleep modes in ultra-dense networks has demonstrated a substantial reduction in EC by selectively turning off BSs during low-traffic periods [27]. However, the main challenge of this method is the need to satisfy the requirements for fair distribution of network resources and precise management of user associations (Figure 4.2), which necessitates further extensive investigations in order to have full implementation of BS sleep techniques in future dense 5G and 6G HetNets.

4.2.7 Sleep Mode Methods Based on Cooperative Communications

Network operating schemes based on cooperative communications in HetNets promote collaboration between macro and small BSs to optimize network energy savings, while ensuring network QoS through effective management of network resources [33]. Thus, network EE can be improved through cooperation among BSs in the effective reallocation of UDs when specific BSs enter sleep mode (Figure 4.2). This can be achieved through the use of different stochastic models for modeling pseudo-randomly changing network states such as the Markov model in [34].

A promising technique for establishing the concept of sleep mode methods based on cooperative communications is coordinated multi-point (CoMP) transmission [29]. CoMP transmission enables multiple BSs to work together to serve UDs simultaneously instead of operating independently, thus improving UD signal quality and reducing BS transmit power needed for coverage. In CoMP transmission, BSs serving the same UDs coordinate their transmission timings and beamforming patterns to reduce interference and optimize resource allocation. Through this BS coordination and cooperation, CoMP enables intelligent traffic redistribution among UDs, allowing some BSs to enter sleep mode while nearby BSs cooperate to maintain coverage. The network can dynamically select the best BS to serve the UD based on real-time conditions, thus reducing redundant transmissions. This prevents overloading and can contribute to the reduction of network power consumption without degrading user experience. Also, instead of frequent handovers between BSs (which consume energy), CoMP enables seamless transitions between coordinated BSs, which can result in fewer dropped connections and reduced energy-intensive signaling overhead.

4.2.8 Sleep Mode Optimization Using Machine Learning Techniques

ML techniques can be integrated with the BS sleep mode approach to optimize EC in mobile networks by enabling intelligent decision-making processes based on variations in traffic patterns and changes in network conditions. By analyzing historical and real-time traffic pattern data, ML algorithms can predict fluctuations in user demand and dynamically adjust the activation and deactivation of BS(s) accordingly (Figure 4.2). These algorithms can classify different traffic scenarios and determine the optimal time to switch BS(s) or BS(s) components into low-power states without compromising network performance.

For example, techniques based on reinforcement learning can continuously improve sleep mode strategies by learning from previous actions and outcomes, ensuring increasingly efficient energy-saving decisions. Deep learning techniques can analyze complex network behaviors and detect usage patterns that traditional rule-based approaches to network resource management might overlook. ML techniques can also facilitate cooperative strategies among BSs, allowing active BS(s) to seamlessly handle traffic loads while neighboring BS(s) enter sleep mode. Predictive analytics of ML techniques can contribute in minimizing delays caused by reactivating BS(s) from sleep mode of operation by anticipating peak network traffic times and ensuring smooth transitions between on/off states of BS(s) or BS(s) components. Unsupervised ML techniques can detect anomalies in network activity and identify inefficient energy usage patterns, which enables further optimization of mobile network power consumption management. Thus, the implementation of different adaptive ML techniques can enable adjustments to evolving network demands, making energy-saving strategies more effective over time. Through continuous learning and adaptation, the ML techniques can enhance the intelligence of sleep mode implementations, significantly reducing power consumption while maintaining network QoS.

4.2.9 Literature Overview of the Recent Implementation of Sleep Mode Methods Based on ML Techniques

Implementing sleep mode methods for entire BS or BS components based on the results of ML processing has been analyzed as the potential approach to significantly improve the EE of current 5G and future 6G mobile networks (Figure 4.2). Table 4.3 presents an overview of recent

TABLE 4.3

Overview of Recent Research on Mobile Network EE Optimization through a Combination of Sleep Techniques and Machine Learning

Reference	Research Focus	Approach/Method	Results
[35]	EC optimization in ultra-dense networks (UDNs) meeting 5G requirements	Machine (reinforcement) learning for BS sleep mode management	EC reduction through deactivation of key BS components including RF PAs and digital signal processing elements; optimized EC based on traffic and latency requirements
[36]	Optimization of EC and energy sharing among small BSs powered by renewable energy sources	Machine learning (Q-learning, deep Q-learning, and imitation learning)	Reduction of grid EC by 31% to 58% in residential zones and by 34% to 68% in business zones; operational cost reduction of up to 35% in residential and 44% in business zones over a 10-year period
[37]	BS EC optimization in 5G networks while maintaining QoS	Q-learning algorithm for implementation of an advanced sleep mode (ASM) strategy	Reduction of EC through gradual deactivation of BS components; determination of optimal values for EC and QoS trade-off; significant energy savings under low load
[38]	Optimization of EE in wireless networks through BS resource management	Machine learning to predict BS on/off states and user-BS associations	The approach saves 99% optimization time; identifies non-critical BSs that can be switched on and off, and manages user associations to enhance network energy efficiency
[39]	Optimization of BS EC with environmental sustainability considerations	Application of machine learning and Grey Wolf Optimizer (GWO) techniques for optimal BS sleep management	Significant BS EC reduction by combining GWO, AI, and sleep procedures; improved EE in wireless networks
[40]	Dynamic multi-sleeping control in 6G networks with diverse QoS requirements	A federated reinforcement learning-based (FRL-based) approach to managing small BS sleep states	Addresses concept drift by adapting to real-time traffic and user mobility patterns; enables high EE while maintaining QoS under ultra-dense network conditions
[41]	Application of machine learning techniques to improve EE in 5G and 6G networks	Review of techniques for network optimization and EC reduction	Identified machine learning methods that significantly contribute to reducing EC and support sustainable 5G and 6G network development

research on mobile network EE optimization that exploits a combination of BS sleep mode methods and ML techniques.

In the context of reducing energy consumption in ultra-dense networks (UDNs) with 5G network requirements, the use of ML techniques has proven to be very effective. In [36], a solution was proposed that uses reinforcement learning to manage the sleep modes of BSs to achieve an optimal balance between energy efficiency and QoS (Table 4.3). It is shown that through different sleep levels, it is possible to turn off key components of BSs, including RF PAs and digital signal processing components (digital baseband), thereby achieving significant energy savings, especially in scenarios with low-traffic load. This approach allows for the adaptation of EC depending on the current traffic and latency requirements, thereby optimizing the operation of BSs in 5G networks.

Additionally, the study in [36] suggests a system for optimizing energy sharing and sleep mode between small BSs that are powered by renewable energy sources (Table 4.3). This system utilizes a variety of ML techniques, including Q-learning, deep Q-learning, and imitation learning. Q-learning is a reinforcement learning method in which an agent learns the optimal policy through interaction with the environment, while deep Q-learning applies deep neural networks to decision-making in complex network environments. Imitation learning allows models to replicate expert decisions. The results show significant network energy savings and a reduction in OPEX of the mobile network operator (Table 4.3), which makes this strategy suitable for future networks with high energy requirements.

Furthermore, in [37] the trade-off between optimization of BS EC in 5G networks and maintaining network QoS has been analyzed through the implementation of advanced BSs sleep modes (Table 4.3). Authors demonstrate in [37] that the Q-learning approach optimizes advanced sleep modes in 5G BSs, allowing for the intelligent deactivation of BS components based on traffic load dynamics (Table 4.3). The proposed approach thus enables a substantial reduction of network EC.

In [38] the authors analyze the possible usage of ML to predict BS on/off states and user-BS associations (Table 4.3). The study in [38] employs neural networks to predict which BSs can be deactivated and which UD connections should be reallocated to other active BSs (Table 4.3). It was demonstrated that this ML method based on neural networks achieves performance that is comparable to methods based on mathematical optimization, while simultaneously reducing computational complexity and execution time.

In [39] optimization of BS EC with environmental sustainability considerations through the utilization of ML algorithms in conjunction with Grey Wolf Optimizer (GWO) techniques has been analyzed (Table 4.3). The presented results demonstrated that combining GWO and AI can be highly effective in the reduction of BS EC when employed for managing BS sleep mode procedures (Table 4.3).

Additionally, a recent study on 6G networks suggests a federated reinforcement learning (FRL)-based approach for BS dynamic multi-sleeping control, which addresses a variety of QoS requirements in ultra-dense networks by enabling small BSs to enter optimal sleep states based on varying traffic fluctuations and user mobility patterns [40] (Table 4.3). Results in [40] show that federated learning has been especially beneficial, because of its capacity to address the issue of concept drift. Concept drift is the phenomenon of model performance degradation which is a consequence of evolving data patterns, such as the relocation of users and the evolution of QoS requirements, that differ from the data on which the model was initially trained. By consistently adapting to these network conditions in real-time, BSs can maintain a high EE without compromising QoS.

A review of recent literature presented in [41] highlights the significant role of ML techniques in improving the EE of BSs, especially through precise control of sleep modes in 5G and future 6G networks (Table 4.3). Approaches that are based on the implementation of reinforcement learning, Q-learning, and neural networks enable dynamic deactivation of BS components in ultra-dense networks, achieving energy savings while maintaining QoS. Furthermore, the integration of ML with optimization methods and renewable energy sources improves network sustainability. Implementation of ML techniques improves the EE of BSs by enabling precise execution of adaptive modes of operation and optimizing resource allocation, even in the complex environments of 5G and future 6G networks [41] (Table 4.3). With intelligent and dynamic mobile network resource management, ML will have a key role in meeting EE requirements for next-generation mobile networks, achieving a balance between expected high mobile network performance and reduced EC.

4.3 Optimization of Base Station Radio Transmission and Resources

Optimizing BS radio transmission and resource utilization is essential for achieving high mobile network EE. The primary goal of these optimizations is to efficiently use the mobile network RF spectrum and network equipment resources in a way that that optimizes the EC of the mobile network. Table 4.4 summarizes key techniques and corresponding features for BS radio transmission and resource optimization that contribute to improving mobile network EE.

Generally, key techniques for BS radio transmission and resource optimization can be broadly divided into the next techniques that contribute to improving mobile network EE (Table 4.4): BS Tx power optimization, optimal

TABLE 4.4

Summary of Key Techniques and Corresponding Features for BS Radio Transmission and Resource Optimization that Contribute to Improving Mobile Network EE

References	Techniques	Key Features
[42–47]	Transmit (Tx) power optimization	Core techniques include BS Tx power control and dynamic adjustment of BS transmission power, resulting in significant energy savings. Advanced approaches use deep learning to predict the signal-to-interference-plus-noise (SINR) ratio
[47–54]	MIMO and mMIMO technology	MIMO uses multiple antennas for simultaneous transmission of multiple signals, increasing network capacity and reducing EC. Transmission power optimization in mMIMO systems is key for energy-efficient 5G networks
[55–61]	Cognitive radio and cooperative techniques	Cognitive radio enables intelligent adjustment of transmission parameters according to spectral conditions. Cooperative relaying facilitates adaptive transmission modes, optimized relay placement, and efficient power allocation to balance EE and QoS, while dynamic resource allocation optimizes spectrum use
[52, 58, 62]	Innovative cloud network architectures	Massively distributed antenna systems with fiber optics and deep learning in mMIMO and heterogeneous cloud RAN (H-CRAN) systems enhance power and resource distribution, reduce interference and EC, and improve EE

implementation of multiple-input multiple-output (MIMO) and massive MIMO (mMIMO) technologies, exploitation of cognitive radio and cooperative techniques among BSs, and modern network implementations based on innovative cloud network architectures. These techniques can contribute to improving network EE when combined with intelligent resource allocation strategies and network configurations that adapt to user requirements and conditions [63]. Each of these techniques is explained in subsequent sections.

4.3.1 Transmit Power Optimization of the Base Stations

EC optimization through BS Tx power control is a fundamental technique for EE optimization within the RAN part of the mobile network. Initially, in mobile networks, this technique is based on static adjustment of BS Tx power according to the size of the BS coverage area, the number of UDs within the BS signal coverage range, wireless connection quality, and environmental conditions impacting wireless signal propagation. Dynamic Tx power adjustment, a more advanced method, can contribute to enhancing network EE by continuously optimizing the BS Tx power level in real time. Figure 4.3 presents BS Tx power optimization concept. This enables adapting in real time the BS Tx power to the current variations in the number of users and network traffic load [42] (Table 4.4). Thus, the BS Tx power optimization can improve mobile network EE through dynamic adjustments of BS Tx power based on traffic demand and network conditions, which contributes to reducing unnecessary network EC that is characteristic for BS transmitting at a constant static Tx power level.

This technique is particularly useful in multi-layer networks and urban areas, where efficient spectrum utilization is critical. In urban environments, where the average distance of a UD from the BS is shorter and the data traffic volume is generally higher when compared with those in rural areas, dynamic BS Tx power optimization is particularly effective (Figure 4.3). It is important to emphasize that solutions based on the Tx power adjustments do not require upgrading the hardware components of the BS, which makes practical implementation of this technique easier. These highlight the key role of BS Tx power management in supporting sustainable and energy-efficient 5G and future 6G networks.

Also, the Tx power control can significantly contribute to reducing the negative impact of interference on mobile network EE. A possible approach to reducing interference in the network is seen in the implementation of techniques that employ dynamic BS Tx power scaling according to space and time variations of UD traffic. This approach can contribute to the reduction of inter-cell interference, while at the same time contributing to the reduction of overall BS power consumption [23]. This is because BS transmission at higher Tx power levels results in higher BS energy consumption and vice versa (Figure 4.3). An approach based on adjusting BS Tx power according to space and time variations of network traffic reduces BS EC in comparison with the traditional approach to BS operation based on transmission at constant Tx power. Thus, the development of adequate interference management strategies based on dynamic Tx power scaling can enable a significant EC reduction of mobile networks, while also reducing network interference.

Some recent research shows that advanced AI-driven approaches, such as deep learning-based BS Tx power control, bring a new dimension to BS Tx power optimization. A study in [43] suggests a mechanism that utilizes a convolutional time-series prediction model to forecast signal-to-interference-plus-noise ratio (SINR) and adjusts BS Tx power to consume minimal energy while preserving the required QoS. This approach has shown significant energy savings under fluctuating network traffic conditions, which is essential for the long-term sustainability of mobile networks.

Furthermore, a study on energy-efficient microcell BS power control in heterogeneous cellular networks presented in [44] employs a stochastic optimization algorithm to analyze the impact of BS Tx power on EE. The proposed optimization algorithm maximizes network EE by determining the optimal transmission power for microcell BSs. In dense HetNets, where high user density and overlapping coverage demand efficient mobile network energy management, simulations highlight the critical role of Tx power control for microcell BSs. The analyzed strategy is in accordance with the principles of dynamic BS Tx power adaptation, which enables microcells to alter power levels in dense environments to conserve energy while preserving QoS.

BS Tx power optimization also plays a crucial role in device-to-device (D2D) communications (Figure 4.2), through intelligent management of BS power levels to enhance mobile network EE. By reducing the Tx power of BS when D2D communication is feasible, UDs can communicate directly with each other rather than relying on BSs, thereby lowering overall network power consumption. Additionally, adaptive BS Tx power management allows the network to dynamically switch between direct D2D links and BS-assisted communication based on real-time conditions, which can further enable reducing the energy consumption of mobile networks. Results presented in [45] show that network energy savings are achieved while maintaining minimal QoS requirements through the application of non-cooperative game theory and fractional programming in optimizing the Tx power allocation in D2D communications of HetNets. This method is seen as essential for the preservation of high performance and EE in future 5G networks.

A side effect of dynamic changes in the BS Tx power and spectrum utilization is that these changes greatly affect the operation of neighboring cells in terms of the BS capacity and throughput. In HetNets, improper BS Tx power distribution can result in increased interference between cells, which reduces network performance (Figure 4.2). Research in [46] highlights the importance of balancing BS Tx power levels to reduce interference while increasing spectral efficiency (SE) and throughput in environments with high traffic demands.

4.3.2 Use of MIMO Technology

MIMO signal transmission technology uses multiple-input and multiple-output antennas and RF chains at both the transmitter and receiver sides to simultaneously transmit multiple independent signal streams over the same RF band. This spatial signal multiplexing technique enables an increase in BS throughput and capacity, without requiring additional bandwidth or BS Tx power.

Besides a significant increase in BS throughput and capacity in comparison with traditional single-input single-output (SISO) technology, MIMO technology also improves BS EE by enhancing BS SE, optimizing BS Tx power allocation, and reducing the time needed for BS transmission at high Tx powers (Table 4.4). A characteristic example of reducing the need for transmission at high BS Tx powers in MIMO communication systems is signal beamforming (Figure 4.4), a key MIMO feature that enables the focusing of the wireless signal energy toward intended UDs. Rather than the signal being broadcast omnidirectionally with significantly higher Tx power, MIMO beamforming contributes to reducing energy wastage and interference.

Additionally, to maintain reliable communication with minimal energy consumption, MIMO also enables adaptive wireless signal transmission

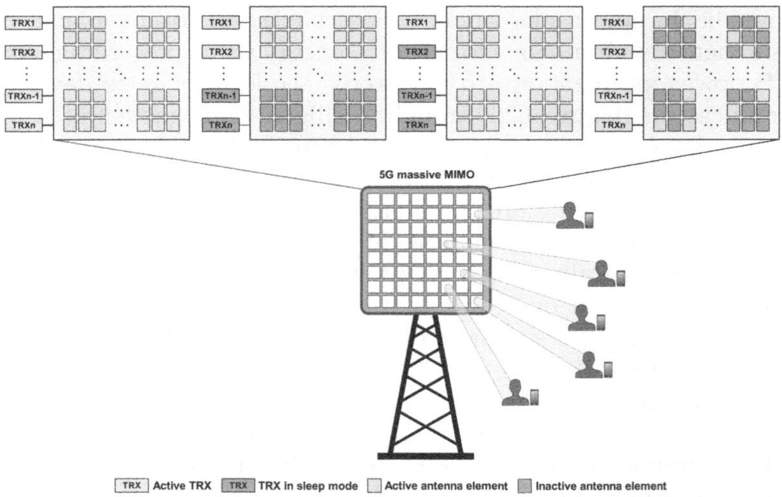

FIGURE 4.4
The massive MIMO technology with different implementation combinations of active transceivers and antenna elements [64]

which dynamically adjusts signal Tx power levels based on wireless channel conditions. By integrating MIMO with sleep mode strategies, BSs during low-traffic periods can shut down unnecessary (MIMO) transceiver (TRX) and antenna components, while still ensuring necessary data rates and coverage to connected UDs. These characteristics of MIMO technology contribute to lowering overall BS power consumption, making MIMO a crucial technology for the realization of energy-efficient mobile networks. Furthermore, the study [48] shows that the combination of standard and cooperative MIMO techniques can enable significant BS energy savings without compromising the performance of the mobile network.

Additionally, the massive MIMO technology represents a major evolution of MIMO technology implemented in 5G networks. Figure 4.4 presents the mMIMO technology with different implementation combinations of active transceivers and antenna elements. By using a large number of antennas, the mMIMO further increases network capacity, making mMIMO technology particularly valuable in high-density and performance-critical usage scenarios. Also, mMIMO systems, in comparison with MIMO systems, further enhance BS EE by improving UD SNR and by allowing BSs to serve more users with lower Tx power per user (Table 4.4). Compared to MIMO technology, spatial multiplexing in mMIMO further increases BS capacity, without the need for increasing the Tx power of BS during transmission. Additionally, new generations of BS Tx power control algorithms developed for mMIMO can dynamically adjust BS Tx power based on real-time channel conditions, ensuring that the PA of BSs consumes energy efficiently, without excessive power consumption that is related to unnecessary signal amplification.

On the other hand, the number of TRXs in mMIMO systems that corresponds to the number of simultaneous signal streams that can be transmitted is significantly larger than in MIMO systems (Figure 4.4). Current typical mMIMO configurations start with 32 × 32 TRXs per BS and can range to 64 × 64, 128 × 128, and 256 × 256 TRXs per BS in 5G systems. More TRXs ensure higher BS capacity but also contribute to an increase in overall BS power consumption, interference, and implementation cost.

As optimizing Tx power allocation and leveraging beamforming techniques in mMIMO systems become essential for enhancing EE in modern 5G networks (Table 4.4), this topic is analyzed in the research papers discussed further. Research in [65] highlights the benefits of mMIMO systems related to minimizing BS power consumption and implementing advanced user scheduling strategies to maximize EE and performance. Further research in [50] concentrates on optimizing BS Tx power in mMIMO systems utilizing imperfect channel state information (CSI) with BS Tx power control. The authors in [50] introduce an improved particle swarm optimization (PSO) technique for regulating user Tx power which enables enhancing BS EE by more than 10% relative to conventional mMIMO transmission approaches. A further study [51] investigates the trade-off between SE and EE in mMIMO systems, emphasizing that energy savings can be realized through the optimization of Tx power and the dynamic adjustment of the number of active antennas according to traffic variations. These adjustments have a minimal impact on the SE, thus preserving the high performance required for 5G networks. In addition, the authors in [47] present a model utilizing zero-forcing (ZF) beamforming technology, which enables modulation of BS Tx power in accordance with channel conditions within the mMIMO systems, facilitating significant energy savings while preserving QoS.

In the context of the application of mMIMO technology in 5G networks, paper [52] presents an innovative approach using deep learning to optimize BS Tx power allocation in mMIMO systems. The proposed method uses deep neural networks to determine the optimal BS Tx power allocation based on UD location, thus achieving efficient and fast power allocation with lower computational complexity. This approach allows for real-time BS Tx power allocation adjustment, which leads to improved EE and network performance, making it crucial for advancing the application of mMIMO technology in 5G networks.

The research in [53] investigates the application of genetic algorithms for the dynamic adjustment of the number of active antennas in relation to the UD number within each cell of mMIMO systems. This technique shows a significant contribution to the reduction of network EC, which is essential for the long-term sustainability of 5G networks employing mMIMO technology. Furthermore, advancements in antenna design significantly influence the development of mMIMO technology. Particularly for millimeter-wave (mmWave) applications, 5G mMIMO and antenna array design have been considerably improved, and this contributes to the improvement of BS EE and network performance [54]. These enhancements are indispensable for meeting the requirements of next-generation networks, as they demand more efficient spectrum utilization and improved adaptability to changing network requirements.

Furthermore, in 6G networks, the number of BS TRXs and individual antenna elements will be even higher than in 5G systems and this pushes the development of MIMO systems toward so-called extreme MIMO (xMIMO) technology. This technology represents an evolution of mMIMO systems that expands the number of TRXs and MIMO antenna arrays to even greater numbers, with the aim of further increasing network capacity, SE, and EE. Compared with mMIMO configurations, xMIMO uses thousands of antennas, which enable greater signal-shaping flexibility, wider coverage, and higher SE under different network loads. By enhancing spatial multiplexing and optimizing BS Tx power allocation, xMIMO will further increase mobile network capacity while reducing the energy cost per transmitted bit. This will make xMIMO an important technology for the development of future sustainable and high-performance mobile network generations. According to preliminary analyses presented in [49], it is expected that the implementation of improved TRX and antenna hardware design with energy-efficient antenna deployment methods ensures that xMIMO systems can maintain high EE, especially in next-generation mobile networks.

4.3.3 Cognitive Radio and Cooperative Resource Allocation Techniques

The fundamental characteristic of cognitive radio (CR) is spectrum sensing capability, which identifies vacant spectrum holes for secondary user devices and uses them to transmit wireless signals without interfering with primary user devices [66]. Figure 4.5 presents the concept of cognitive communications in different mobile network RAN architectures. The CR is an advanced wireless technology that allows network infrastructure elements, such as BSs, to intelligently adjust their activity and Tx parameters based on the current spectrum availability [67–71]. This capability enables more efficient use of the available spectrum, increases network flexibility, and reduces network EC by dynamically adjusting BS and UD Tx power and transmission frequency in accordance with the changing availability of wireless channels (Table 4.4). Additionally, CR facilitates opportunistic spectrum access, allowing BSs and UDs to transmit on the most energy-efficient frequency bands, thereby optimizing signal propagation and reducing Tx power demands. Interference management through CR can also contribute to minimizing retransmissions and packet losses in mobile networks, which can further contribute to conserving network energy by reducing redundant data transmission. CR is thus becoming an important technology in the development of advanced mobile networks, especially those operating in saturated spectrum conditions, since

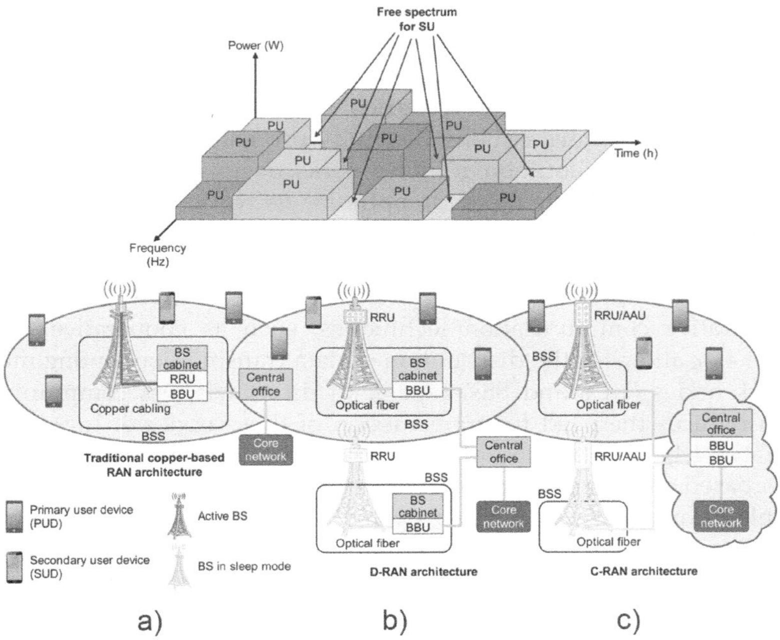

FIGURE 4.5
Concept of cognitive communications in different mobile network RAN architectures, (a) traditional RAN architecture, (b) distributed – RAN (D-RAN) architecture, (c) cloud – RAN (C-RAN) architecture [74, 75]

it enables optimal allocation of spectral resources and improved EE [55, 70, 72] (Figure 4.5). However, optimizing the trade-off between network SE and EE is a continuous challenge in the practical implementation of CR and will be essential for achieving high-performance modern 5G and 6G HetNets.

Different research papers analyze the possible impact of the implementation of CR on the improvement of mobile network EE. The authors in [59] propose a superior population generation algorithm (SPGA) for optimal resource allocation, which enables a balance between SE and EE. Unlike conventional solutions that focus solely on improving SE that often leads to increased mobile network EC and OPEX, the proposed SPGA addresses multiple objectives simultaneously. Compared to conventional methods and regular genetic algorithms (GA), SPGA achieves a better resource allocation with a 10% improvement in network EE at the same level of SE. These results demonstrate the capabilities of SPGAs to enable more sustainable and efficient 5G HetNets, thus offering a practical roadmap and insightful analysis for the advancement of spectrum-aware and energy-efficient mobile network designs.

The study in [60] investigates the adaptation of Tx power in cognitive mobile networks with multihop data transmission. The authors analyze different techniques for optimizing energy consumption, including the Tx power control, frequency allocation, and link scheduling. The proposed algorithms allow for real-time adaptation of network base stations Tx power, by taking into account the complexity of the network environment, which is crucial for improving EE in scenarios with multiple hops between communicating nodes. Furthermore, the study presented in [61] builds on the concept of resource allocation in CR networks, by introducing Tx power control methods using game theory to optimize network SE and EE. The work emphasizes the importance of balancing the Tx power distribution to reduce network interference and achieve energy savings, especially in competitive network environments where spectral resources are shared.

Cooperative communication techniques, such as cooperative relaying (Figure 4.2), allow for the distribution of data transmission among multiple network nodes (UDs and BSs) capable of direct wireless communication, which reduces the need for transmission of those nodes at high Tx powers (Table 4.4). This method allows UDs to share resources and adjust Tx power based on their distance from other UDs and BSs, thus achieving a sustainable balance between EE and QoS [57]. Cooperative relaying improves the EE of mobile networks by enabling relay nodes (UDs or BSs) to assist in forwarding data, which contributes to reducing the Tx power of the BSs or UDs needed for ensuring wireless connection. By utilizing spatial diversity, cooperative relaying minimizes wireless signal path loss and attenuation, which contributes to ensuring reliable communication with a lower power consumption of network elements (BSs and UDs).

The widely used amplify-and-forward (AF) and decode-and-forward (DF) relaying techniques allow intermediate nodes to either boost or regenerate signals, which reduces the number of retransmissions in communication between UDs and BSs, and consequently can contribute to the improvement of network SE. The cooperative relaying is particularly beneficial in edge and cell-edge use scenarios, where direct communication with the BS generally requires higher Tx power on both the BS and UD sides (Figure 4.2). Additionally, relay selection algorithms can ensure that the most energy-efficient relay nodes are used, optimizing power distribution across the network.

Furthermore, effective planning of relay placement and power allocation is crucial to maximize energy efficiency while ensuring compliance with QoS constraints. As shown in [57], adaptive transmission mode selection, including direct, relay, and cooperative communication, combined with optimized relay placement and Tx power allocation, increases network capacity while reducing overall EC. The study further emphasizes the need for careful balancing of relay location and Tx power, as improper placement can increase transmission distances and lead to higher EC.

When combined with BS sleep mode techniques, cooperative relaying can offload traffic from BSs, allowing some BSs to enter low-power states while maintaining network coverage. ML-driven optimization of relay node selection and operation represents a possible solution to further enhance network

EE by dynamically selecting the most efficient relay paths based on real-time network conditions. By reducing transmission distances, mitigating interference, and optimizing Tx power allocation, it is expected that cooperative relaying will in the future have a more dominant role in enhancing the sustainability of 5G and future 6G networks.

4.3.4 Innovative Cloud-Based RAN Architectures

Figure 4.5 presents different mobile network RAN architectures. Contemporary cloud radio access network (C-RAN) is an advanced mobile network architecture that centralizes the processing of BS baseband functions in centralized BBUs located in the cloud-based data center(s). The centralized BBUs are connected via high-speed fiber-optic links with geographically distributed RRHs/RRUs located at cell site locations (Figure 4.5). There are multiple reasons why in the future massive implementation of C-RAN architecture in 5G and 6G networks can significantly contribute to the improvement of mobile network EE.

The first is related to the centralized signal processing that is characteristic of such innovative cloud-based RAN architecture (Table 4.4). The centralized processing and resource optimization performed in BBUs located in the cloud data center(s) reduces RAN power consumption because network architecture based on dislocated individual RRHs/RRUs requires less energy for operation in comparison with traditional BS site locations. Also, centralized processing enables dynamic mobile network resource allocation, ensuring that the computing power of cloud data centers is used efficiently, reducing overall mobile network energy waste that is characteristic of data and signal processing at individual BBUs on dispersed BSSs (Figure 4.5). Additionally, fiber-optic links between distributed RRHs and centralized cloud-based BBUs provide low-latency and high-bandwidth connectivity, which enables energy-efficient data transmission with low signal loss. This is due to characteristics of the fiber optics, which in comparison with copper-based or microwave fronthaul transmission methods require less power per transmitted bit, and this consequently contributes to the improvement of network EE. Also, connecting RRUs/RRHs and centralized BBUs by means of passive optical networks (PONs) composed mostly of passive optical network components can contribute to further reducing network EC in the fronthaul part of the network.

Also, with centralized intelligence of C-RAN architectures, it is easier to implement sleep mode or Tx power scaling techniques that enable dynamic activation or deactivation of RRHs or scaling its Tx power based on real-time traffic demand (Figure 4.5). Implementation of these techniques in C-RAN networks can be particularly beneficial in bringing significant power savings to HetNets. Also, C-RAN supports the virtualization of BBU processing through the network function virtualization (NFV) concept, enabling the sharing of computational resources among multiple RRU/RRH and corresponding cells (Figure 4.5). This reduces the need for redundant servers (hardware) in the data center, leading to lower overall network power consumption. Additionally, the implementation of cloud-based AI and ML algorithms can further contribute to optimizing the power consumption of RRUs/RRHs and centralized BBUs according to network traffic patterns.

Finally, an important advantage of C-RAN architecture is the reduction of energy needed for temperature regulation of traditional macro BSSs, which generally require energy-intensive cooling systems due to the high heat generation of equipment located at BSSs. In the C-RAN architecture, BBUs are located in centralized data centers, where cooling efficiency is higher and waste heat can be managed more effectively (Figure 4.5). Also, RRUs/RRHs mostly rely on natural air cooling, which eliminates the need for additional EC used in active cooling.

Despite many advantages that C-RAN network architecture can bring to the improvement of mobile network EE, there are issues related to expensive and difficult deployment, latency constraints, and data security and privacy constraints. Also, the full potential of C-RAN architecture in enhancing mobile network EE has yet to be fully explored.

Some research attempts are dedicated to the implementation of advanced ML techniques proposed in [58] to improve EE, by optimizing resource allocation within 5G heterogeneous cloud radio access networks (H-CRANs). The study utilizes a centralized resource management strategy to enhance the allocation of BS resource blocks and Tx power distribution, thereby minimizing inter-tier interference and improving network EE while preserving QoS.

4.4 Energy Consumption Optimization in the Backhaul Part of the Mobile Network

Although the RAN part has the highest contribution to overall mobile network EC, the EE of data transmission between the RAN part and the core network (CN) part (known as backhaul) in traditional network architectures also has an important impact on overall mobile network EE. The optimization of data transmission EE between the RAN and core segments of a mobile network can be accomplished by implementing network architectures based on connecting the RRUs/RRHs and active antenna units (AAUs) of a BS to the BBUs and the CN segment through optical fiber technology [74]. Figure 4.6 presents mobile 5G network architecture with the latest generations of macro BSSs containing RRU/RRH and AAU elements. As explained in the previous subsection, this type of connectivity takes advantage of the high bandwidth and low loss of optical fiber links, which increases network SE and reduces overall network EC. Optical connectivity often reduces the need

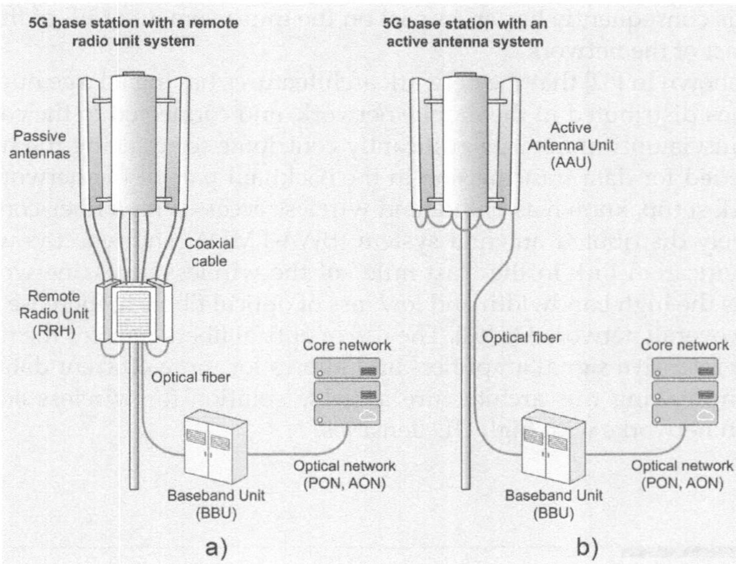

FIGURE 4.6
Mobile 5G network architecture with the latest generations of macro base station sites [77], containing (a) RRUs/RRHs and passive antennas and (b) AAUs

for energy-intensive signal amplification that is characteristic of copper-based, microwave, or satellite backhaul links, resulting in more sustainable and efficient data transmission, especially in modern network environments with dense allocation of UDs.

Additionally, the possible deployment of PONs for connecting RAN and core parts of the network eliminates the need for a significant number of active electronic components in the backhaul network and this contributes to reducing backhaul network EC (Figure 4.6). Also, using different wavelength division multiplexing (WDM) techniques that allow transmission of multiple data streams over a single fiber link and implementing optical switching instead of electronic switching can contribute to lowering energy usage of the backhaul part of the mobile networks [76].

The improvement of the backhaul part of mobile networks can also be achieved through the development of efficient traffic aggregation protocols, data compression algorithms, and implementation of edge caching and content delivery networks (CDNs). Traffic aggregation protocols combine multiple small data flows into fewer, larger packets and thus reduce backhaul network processing overhead, which consequently has a positive impact on the EC of the backhaul network part. Data compression algorithms reduce the volume of transmitted data on the backhaul links and this can contribute to cutting energy use of network elements among which backhaul links are established. Leveraging the fog and edge computing concept, edge caching and CDNs enable storing frequently accessed content closer to users. Such concepts minimize the length of data transfer through a backhaul network and this consequently has an impact on the improvement of EE of the backhaul part of the network.

It is shown in [62] that the network architectures having a large number of antennas distributed in the mobile network, and connected to the core control unit via optical fibers, significantly contribute to reducing the network EC needed for data transmission in the backhaul part of the network. This network setup, known as broadband wireless access with a fiber-connected massively distributed antenna system (BWA-FMDA) shortens the wireless communication link in the "last mile" of the wireless access network and exploits the high bandwidth and low loss of optical fibers to increase SE and reduce overall network EC [62]. The use of optical fibers reduces the need for energy-intensive signal amplifiers and allows for more efficient data transmission, making this architecture a viable solution for wireless access in modern networks with high UD densities.

4.5 Energy Consumption Optimization Using Small Cell BSs

The deployment of small cell BSs (such as microcells, picocells, and femtocells) in creating HetNets has emerged as a prominent strategy for reducing EC in modern mobile networks of 5G and future 6G (Figure 4.2). Small cell BSs allow mobile signal coverage in smaller geographic areas mostly characterized with high traffic demands. Dense allocation of small cell BSs enables UDs to be closer to BSs, which consequently reduces the need for high Tx power of macro BSs. This reduction of the average level of Tx power of BSs during operation contributes to the improvement of mobile network EE [78]. However, the approach of extending network coverage and capacity by installing small cell BSs not only improves network EE but also makes better use of available mobile network resources. Also, UDs located closer to small BSs consume less power to transmit the signal (Figure 4.2), which further extends the UDs' battery life [79]. Small cell BSs thus deployed in network traffic hotspots bring benefits to mobile network service providers and users, in terms that providers can avoid the high costs of dense area coverage with large and expensive macrocell BSs, while users gain faster connections and longer battery life on their devices due to shortening of communication links having better wireless signal quality.

Different research efforts dedicated to improving mobile network EE with the implementation of small cell BSs in HetNets have been presented in the literature. The study in [80] highlights that HetNets realized with small cell BSs can enable significant network energy savings through the implementation of advanced network resource management strategies, which include the utilization of the BS sleep modes and the partial shutdown of small cell BS components when not in use. Additionally, HetNets that combine macro and small cell BSs can further reduce EC by dynamically allocating resources according to changes in user traffic [31]. This approach to deploying HetNets is particularly relevant in the context of the continuously increasing density of network infrastructure elements (BSs or RRUs/RRHs) and the growing demand for more energy-efficient 5G networks.

Furthermore, [81] proposes an EE framework for Internet of Things (IoT) systems based on heterogeneous small cell networks that uses cell-zooming power adaptation algorithms to reduce network EC. This technique allows for significant improvements in the overall network EE, especially in scenarios with high traffic and increased data processing requirements. By leveraging cell-zooming, small cells can efficiently respond to the energy challenges posed by the deployment of 5G and IoT networks.

Additionally, [82] examines energy savings in ultra-dense small cell mobile networks, which manage to achieve EE improvements of up to 70.7% by combining small cell BS Tx power control and UD scheduling within the network. The study [82] highlights that efficient BS resource management within small cell networks can significantly reduce EC and increase overall network reliability, which is crucial for the successful deployment of 5G and future 6G technologies in urban and densely populated areas.

However, along with the development of techniques that can control and reduce the Tx power of small cell BSs, deployment of small cell BSs also brings challenges related to the increased computing power requirements at BSs, especially when advanced technologies such as mMIMO and millimeter-wave communication technologies are implemented. According to research [83], computing power can represent over 50% of the total power consumption of small cell BSs, which highlights the need to optimize not only Tx but also the computing power of BSs to achieve maximum EE.

Moreover, the massive introduction of small cells into 5G HetNets also brings certain practical issues, primarily related to eliminating inter-cell interference among BSs and the implementation of smart BS resource management. For example, research [84] shows that the introduction of small cell BSs in mobile networks can lead to the increased complexity of network management due to the occurrence of mutual BS interference. Reducing the impact of interference requires the application of advanced techniques such as interference coordination and the development of the SONs. Solving these challenges requires careful planning and implementation of small cell BSs to ensure not only network EE but also network operating stability and QoS. In addition, [85] highlights the importance of using different techniques to improve the EE of small cell HetNets. These techniques include the implementation of a sleep (idle) mode of BS operation, energy harvesting, and traffic load balancing, while simultaneously solving challenges related to the efficient use of network spatial resources. The combination of these approaches provides a comprehensive framework for optimizing the EE of small cell BSs in complex network environments.

However, the implementation of small cell BSs in the realization of HetNets has drawbacks, which are primarily related to the contribution of small cell BS EC to overall mobile network EC. This drawback will especially escalate in the future, since the number of installed small cell BSs in 5G networks and especially 6G networks will be significantly higher than the number of installed macrocell BS. This will result in a significant contribution of small cell BSs to the overall network EC of the future HetNets. As a consequence, it can be expected that the overall network EC of mobile networks on a national or global level will continue to increase [77]. Thus, research efforts related to the development of new approaches that will exploit the benefits of small cell BSs while also optimizing mobile network EE, will need to continue.

4.6 Implementation of Renewable Energy Sources

Individual macro BSSs, especially those hosting multiple mobile network technologies such as the 2G, 3G, 4G, 5G, or a combination of some of these technologies represents high-power consumers. Instantaneous power consumption of such BSSs can typically span from a few up to 20 kW. In addition, some remote BS locations face a unique challenge when it comes to power supply, as they sometimes cannot connect directly to the main power grid supply. Instead, such off-grid BSS locations usually rely on diesel generators that require a regular supply of fuel, which makes this approach the least energy-efficient and most environmentally harmful solution for powering BSSs [86].

As a solution for the power supply of such BSSs, the concept of powering BSSs with energy from renewable energy sources is being implemented in practice [86, 87]. The inclusion of renewable energy sources, such as geothermal, solar, or wind energy, and a combination of these energy sources for the power supply of BSSs is a key step toward reducing reliance on traditional fossil fuels and decreasing overall EC in wireless communication networks. Also, some telecom operators use hydrogen energy produced from renewable energy sources for the remote, off-grid, or backup BSS power supply, thus replacing or supplementing traditional power supply from diesel generators. BSSs powered by renewable energy can significantly reduce greenhouse gas (GHG) emissions, thereby advancing United Nations sustainable development goals and reducing the environmental footprint of the telecommunications industry [88].

Solar energy is the most common type of renewable energy source for the power supply of BSSs, since it is especially valuable in remote areas having appropriate solar irradiation that lack access to the electricity grid, or where the electricity grid connection is limited or expensive. Wind energy is also a common source of energy for powering the BSSs, since wind turbines can offer solutions for power generation in regions that have favorable conditions for using wind as a renewable energy source. However, the nature of renewable energy sources means that they are not always reliable, as they depend on weather conditions. This requires the use of high-capacity batteries at the BSSs to store excess energy and use the stored energy during periods when there is no renewable energy production. Besides higher capital expenditures (CAPEX) related to the implementation of energy storage batteries and renewable energy sources in BSS locations, these implementations create challenges for maintaining the continuity of power supply to BSSs using renewable energy sources [4]. When there is a shortage of energy from renewable energy sources or from energy storage batteries, and if BSSs lack any other source of energy, BSSs stop working, which can negatively affect network performance and user satisfaction. Therefore, proper planning of renewable energy production capacities, as well as energy storage resources, is crucial for the operation of renewable energy-powered BSSs.

4.6.1 Energy Harvesting Technologies

The four primary sources of energy harvesting that can be used to power the mobile network infrastructure and UDs in wireless communication systems are shown in Figure 4.7. These sources include solar or light energy, thermoelectric conversion, electromagnetic radiation, and mechanical motion or vibration, all of which can be used to harvest energy for power supply in wireless communication systems [89].

To address the challenges of powering remote BSSs with renewable energy sources, various energy harvesting techniques have been proposed. Energy can be harvested from single or multiple energy sources including renewable energy sources such as solar (Figure 4.7) and wind energy. The harvested energy can then be used to power conventional BSSs, reducing the dependence on more expensive and less environmentally sustainable energy coming from traditional energy sources.

Solar or light energy is generated by converting sunlight or artificial indoor light into electrical energy using the photovoltaic effect. Although outdoor implementations based on converting sunlight for powering BSSs benefit from abundant sunlight, sunlight availability depends on the time of day and atmospheric conditions. In contrast, indoor lighting offers a more consistent but weaker source of energy that is suitable for powering smaller UDs.

Mechanical energy from motion and vibration can be harvested by electrostatic, piezoelectric, electromagnetic, and wind turbine-based methods. Wind turbines of appropriate capacity harness the mechanical energy of motion generated by wind to generate electricity, offering a suitable solution for the power supply of BSSs.

On the other hand, electrostatic, piezoelectric, thermoelectric, and electromagnetic power-supply methods are more convenient for powering UDs that can span from cell phones to IoT devices. Electrostatic harvesting exploits

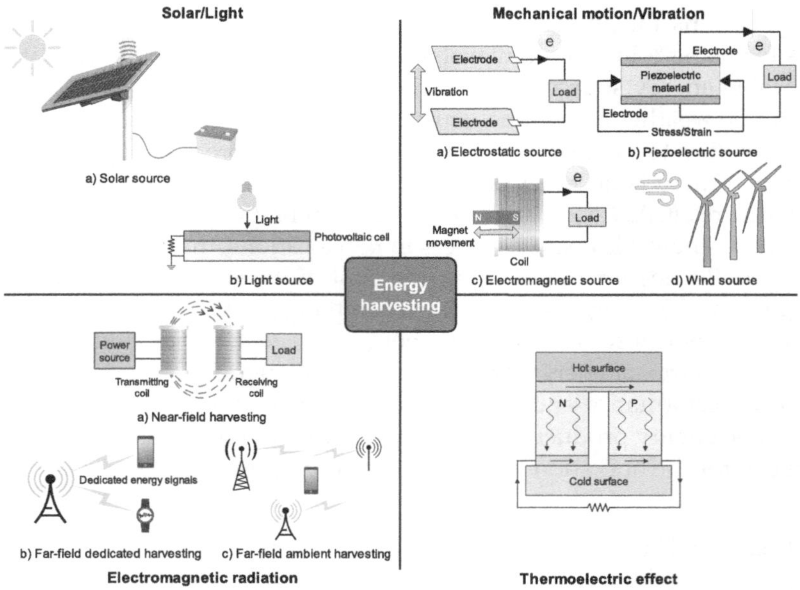

FIGURE 4.7
Energy harvesting sources in wireless communication systems [89]

changes in capacitance caused by motion, while piezoelectric harvesting converts mechanical stress into electrical energy. Electromagnetic harvesting generates electrical current through the relative motion of magnets and coils. Electromagnetic radiated energy harvesting is a method that captures RF signals emitted by sources such as BSs, television antennas, or WiFi transmitters. This can be achieved by near-field methods, which involve the use of magnetic coupling or resonance to transfer energy, or by far-field methods, which involve the conversion of ambient electromagnetic radiation using rectifiers and antennas. Thermoelectric energy harvesting is based on the thermoelectric effect, which generates electrical energy by exploiting temperature differences. This method is particularly useful for applications that involve thermal gradients, such as wearable UDs that generate energy from body heat that is exposed to cooler ambient air.

Together, these technologies can facilitate the development of the UD or BS power supply that can be self-sufficient and thus can contribute to significantly reducing their dependence on conventional energy sources. Therefore, these technologies can support the development of efficient and sustainable wireless communication systems, especially in future network environments that are expected to be densely populated and highly demanding. However, further research activities are needed to reach the point when the implementation of energy harvesting technology can achieve large-scale deployment in the powering of wireless UDs and BSSs.

4.6.2 Application of Energy Harvesting in 5G and IoT Networks

The increasing energy demands of UDs in 5G networks and the rapid growth of IoT devices have made energy harvesting and management technologies essential for their massive implementation. Energy harvesting technologies aim to reduce dependence on conventional energy sources, improve network sustainability, and extend the lifetime of IoT devices, especially in resource-constrained scenarios.

One of the key techniques that leverage RF energy harvesting for UDs is simultaneous wireless information and power transfer (SWIPT), which enables devices such as wireless sensors, low-power IoT devices, and mobile terminals to receive data while concurrently harvesting energy from RF signals [90]. When a SWIPT-enabled BS or access point (AP) transmits an RF signal, the receiver device extracts useful energy from the RF signal to power itself and decode information embedded in the RF signal for communication. This process requires specialized RF circuit design that can separate and optimize energy harvesting and data decoding. However, SWIPT is particularly interesting in mMIMO implementations where part of RF transmission from some antenna elements can be dedicated to data transfer, while another part can be dedicated to energy harvesting. Also, SWIPT technology shows a promising implementation in mmWave and terahertz (THz) frequencies, which have high throughput availability.

Thus, SWIPT is particularly valuable for the power supply of battery-free or energy-constrained devices, since SWIPT technology supports green communication initiatives by minimizing dependence on traditional power sources and extending device lifespan. This is beneficial in dense network environments typical for 5G networks and beyond. SWIPT therefore represents an important step forward in the development of self-sustaining, energy-efficient communication systems that support current environmental goals. However, massive implementation of SWIPT technology in UDs is still in its infancy and additional research needs to be done in the development of receiver circuit design that will cope with SWIPT low RF energy conversion efficiency, interference between energy and data signals, transmission range limitations, and security and privacy issues.

In addition, authors in [91] propose a software-defined energy harvesting network (SD-EHN) architecture that uses the software-defined networking (SDN) concept to dynamically manage data and energy flows within 5G networks, enabling efficient use of renewable energy sources such as solar and wind energy. It is shown that the implementation of SD-EHN reduces dependence on conventional energy sources and GHG emissions.

In response to the increasing demand for EE and SE in 5G networks, the authors in [92] propose an integrated approach to spectrum management and energy harvesting. This strategy enables the coordinated use of energy and spectrum resources, which is essential for creating sustainable and energy-efficient communication networks while reducing dependence on traditional energy sources. The proposed energy and spectrum management mechanisms enable dynamic management of resources in different network layers, including D2D networks, small cell networks, and macro cell networks, thereby significantly improving the overall EE of 5G networks.

Furthermore, the authors in [93] investigate various energy harvesting techniques, including solar, electromagnetic, thermal, and mechanical sources, and their application in IoT devices. These techniques enable self-sustaining networks that can operate without frequent battery replacements, which is crucial for the long-term sustainable management of IoT devices in 5G systems. The integration of these technologies supports the goals of reducing the environmental footprint and enables reliable and durable communication networks, making them a key part of strategies for future green communications.

Additionally, the authors in [94] further investigate the application of energy harvesting techniques in the IoT ecosystem, highlighting their key role in preserving wireless sensors' operation and extending their battery lifetime. The authors analyzed different energy sources for the power supply of IoT devices, such as solar and RF energy, and showed how techniques of solar and RF energy harvesting can replace or complement conventional battery-based power supply systems. In addition, the study provides case studies of IoT devices using energy harvesting techniques, addressing integration challenges, and suggesting necessary optimizations to reduce the EC of IoT devices. The presented improvements form the basis for the development of sustainable, energy-efficient IoT networks, especially in the context of the rapid expansion of 5G networks and IoT applications.

4.6.3 Capacity Planning for Solar and Wind Power Systems

Effective capacity planning of solar and wind power plants with integrated battery storage systems is essential for ensuring the continuity of power supply at remote BSS locations. It is important to properly size these systems to ensure a reliable power supply, given the unpredictability and intermittent nature of renewable energy sources. Battery systems allow for the storage of excess energy during periods of high production, providing power during periods of reduced production.

The authors in [95] explore optimization strategies for the implementation of hybrid renewable energy systems for the power supply of BSSs, that includes photovoltaic systems, wind turbines, and energy storage batteries. The presented results highlight the importance of accurate dimensioning and capacity management of energy production and storage elements to increase reliability and reduce OPEX of BSSs. In particular, the need for careful BSS battery capacity planning is emphasized to mitigate supply fluctuations and ensure a stable power supply of BSSs, even during periods of reduced renewable energy production.

In line with these findings, the study in [96] provides an overview of the design and deployment of solar-powered BSSs, focusing on challenges related to dimensioning photovoltaic panels and storage batteries. The study underscores the importance of optimal component sizing to minimize power interruptions and maintain a reliable power supply of BSSs, even during low sunlight periods or adverse weather conditions. Additionally, authors in [96] also propose different system configurations, including a combination of solar panels and diesel generators, which can provide redundancy and reduce the risk of power outages, especially in regions with limited access to the grid.

Furthermore, [97] provides a comprehensive modeling analysis of power supply systems based on solar and wind energy sources with integrated battery storage. Special emphasis is given to the challenges of the intermittent nature of renewable energy sources and the need for sufficient battery capacity to ensure continuous power supply of BSSs. The research also highlights the advantages of using hybrid power supply systems that combine multiple energy sources in rural and hard-to-reach areas without access to the electricity grid. This also contributes to further increasing the stability of the BSS power supply.

In addition to the presented challenges in implementing renewable energy sources for the power supply of BSSs, research in [98] analyzes the technical feasibility of a hybrid system that combines solar and wind power with an adiabatic compressed air energy storage (CAES) system to power remote BSSs. The authors emphasize the importance of properly sizing photovoltaic panels, wind turbines, and storage capacities, including the air storage system, to ensure a reliable power supply and minimize the risk of power outages. The results show that configurations with a larger number of photovoltaic panels or increased air storage capacity significantly enhance the system's ability to store and supply energy during low production periods, thereby improving overall BSS power supply stability.

4.7 Application of NOMA Technique for Enhancing Energy Efficiency in Mobile Networks

Unlike traditional orthogonal multiple access methods (such as orthogonal frequency division multiple access [OFDMA] utilized in 4G and 5G networks), the non-orthogonal multiple access (NOMA) technique allows multiple UDs to share the same time, frequency, and spatial resources while separating UDs based on different Tx power levels or code sequences. Figure 4.8 presents a comparison of the Tx power allocation in orthogonal multiple access and NOMA systems. As explained in Chapter 2, Section 2.3.3,

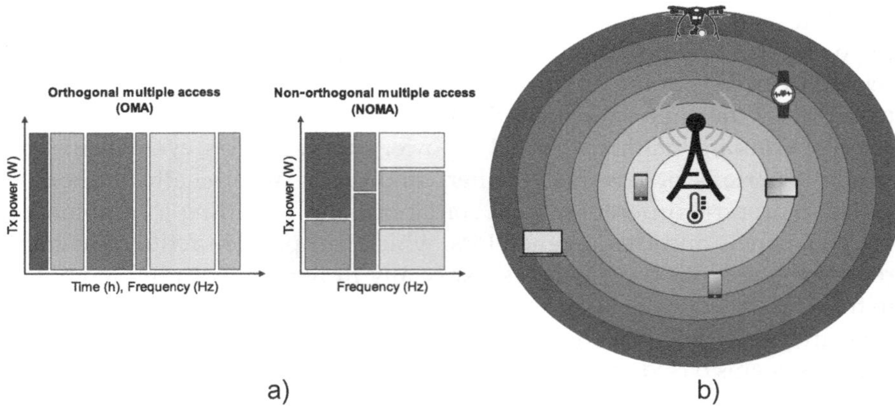

FIGURE 4.8
Comparison of the transmit (Tx) power allocation in (a) orthogonal multiple access and (b) NOMA systems

the NOMA technique is based on the transmission of stronger signals (at higher BS Tx power) to users located farther from the BS, while weaker signals (transmitted at lower BS Tx power) are assigned to users closer to the BS. NOMA also uses successive interference cancellation (SIC) at the receiver side, which allows efficient signal decoding without increasing the Tx power of the transmitting side.

NOMA is thus an advanced mobile network radio access technique that contributes to improving network EE and SE. In particular, the NOMA technique can contribute to reducing the overall mobile network EC, since it exploits the flexible adjusting of the Tx power of BSs according to the channel conditions of each UD [99] (Figure 4.8). Such an approach reduces overall BS power consumption by transmitting fewer independent signals toward served UDs, which reduces the BS transmission duration per UD and minimizes unnecessary retransmissions, all of which contribute to lowering BS EC.

Also, NOMA improves SE by allowing more users to share the same frequency RB simultaneously. This maximizes frequency spectrum usage in mobile networks, which reduces the need for extra frequency RBs. In ultra-dense HetNets, this allows serving more UDs per frequency channel, resulting in fewer active BSs needed to handle the same traffic and, consequently, this contributes to the improvement of the EE of mobile networks. Additionally, advanced BS Tx power control and SIC techniques can reduce interference among UDs sharing the same spectrum, which leads to significant improvements in network EE through optimized resource utilization [100].

Furthermore, NOMA can contribute to the improvement of EE on the UD side. In different multiple access schemes utilized in previous generations of mobile networks (2G–5G), each UD waits for its allocated time, code, or frequency slot, leading to higher EC of UD due to frequent operation in an active monitoring state (Figure 4.8). Implementation of the NOMA technique is based on UDs receiving wireless signals simultaneously, which results in reduced time spent by the UD in energy-intensive active monitoring mode. This will contribute to extending the battery life of mobile UDs and IoT sensors and reducing the UD power consumption during multi-user communications.

Besides the described advantages that NOMA can bring in terms of improving mobile networks or UD EE, for the full implementation of NOMA in practice different challenges need to be solved. One of the challenges is related to reducing the high computational complexity of NOMA receivers based on utilizing SIC, which increases processing power demand and thus contributes to an increase in power consumption. Another issue characteristic of NOMA technology is interference management, where more UDs sharing the same spectrum means higher signal interference, which will require the development of optimized decoding techniques that also need to be energy efficient. Additionally, an important challenge is the development of an efficient BS Tx power allocation of the NOMA technique, since successful implementation of the NOMA technique will require dynamic BS Tx power control to balance EE and signal quality. Some of the recent research activities, described further, tend to contribute to solving some of the stated challenges.

For example, in [101] resource optimization for NOMA HetNets to achieve higher EE in the context of 5G communication systems is investigated. The authors propose approaches for resource allocation and downlink BS Tx power optimization in NOMA HetNets, including UD scheduling and BS Tx power allocation methods. The findings show that these proposed approaches can significantly improve the EE of NOMA HetNets compared to traditional methods such as OFDMA systems. The study [101] also highlights how NOMA can improve network EE by reducing user interference and optimizing BS Tx power allocation in networks with heterogeneous architectures.

Furthermore, the application of NOMA in mobile networks increases EE through dynamic resource optimization [102]. In [102], subchannel assignment and power distribution algorithms in NOMA downlink are proposed, considering QoS requirements, channel scheduling stability, and BS Tx power transmission constraints. Simulations show that these algorithms can significantly improve EE and balance traffic latency. This provides guidance for the implementation of different energy-efficient network management strategies in 5G networks, while at the same time laying the foundations for resource optimization and strategy development applicable in future 6G networks [102].

As analyzed in [103], the combination of the NOMA technique and SWIPT technology offers a promising approach to improve EE in next-generation wireless networks. The study introduces an optimization framework that jointly deals with BS Tx power distribution and time-switching (TS), with the goal of maximizing EE while ensuring compliance with critical constraints such as BS Tx power limits, minimum data rate requirements, and energy harvesting thresholds for each UD terminal. Simulation results show that combining SWIPT with NOMA technologies achieves significant gains in network EE compared to traditional multiple (orthogonal) access systems and highlights the potential of implementing this approach for realizing energy-efficient wireless communication [103].

Energy harvesting technology has the potential to enable the power supply of UDs located near BS(s) with transmitted wireless signal energy of BS(s), while simultaneously enabling an exchange of information between BSs and UDs. However, the high power supply requirements for wireless energy needed to be delivered to UDs, which by far exceed those required for UD operation and information transmission, pose significant challenges in the practical realization of such systems. To address these challenges, NOMA emerges as a promising solution to increase network throughput, while simultaneously meeting energy harvesting requirements as shown in [104]. This study presented in [104] introduces a transmit-TS protocol that enables simultaneous energy harvesting and information decoding. Using path-following algorithms, the research in [104] addresses the key problems of network throughput optimization and EE maximization under power and energy harvesting constraints. Additionally, the performance of the proposed transmit-TS protocol is compared with that based on conventional power-sharing (PS) protocols. Numerical results show that transmit-TS algorithms achieve superior throughput and EE, providing a compelling argument for their adoption in networks that utilize energy harvesting [104].

Another method that can achieve significant improvements in mobile network EE is the use of deep learning techniques, especially in optimizing energy allocation in MIMO-NOMA communication systems. The application of deep learning techniques to optimize EE in MIMO-NOMA systems marks a significant step forward in the design of energy-efficient wireless networks. In [105], an approach based on convolutional deep neural networks (CDNNs) was proposed to solve the BS Tx power allocation optimization problem in MIMO-NOMA systems. CDNN uses advanced deep neural network architectures to efficiently manage resource allocation, overcoming the limitations of traditional optimization approaches. Simulation results show that the CDNN framework achieves superior performance, which is reflected in a better balance between network EC and throughput in comparison to conventional optimization methods [105].

Furthermore, advanced deep reinforcement learning (DRL) techniques offer significant potential for optimizing resource allocation in uplink NOMA systems. In [106], a two-step DRL framework is proposed to jointly optimize subcarrier assignment and Tx power allocation in uplink multi-user NOMA systems. The framework aims to maximize the long-term EE while ensuring the minimum data rate requirements for all UDs. The approach combines a deep Q-network (DQN) for subcarrier assignment with a deep deterministic policy gradient (DDPG) approach for Tx power allocation, ensuring effective handling of both discrete and continuous optimization variables. Simulation results demonstrate that the proposed DRL framework achieves increased EE compared to conventional resource allocation methods, particularly under varying Tx power constraints, thereby enhancing overall network performance [106].

In conclusion, the NOMA technique shows strong potential for improving EE in 5G and next-generation mobile networks. By allowing the sharing of the same frequency resources among users with optimized BS or UD Tx power, NOMA can significantly reduce EC in networks with high data traffic demands. Despite currently not being fully standardized, preliminary research shows that integrating NOMA with technologies such as SWIPT, mMIMO, energy harvesting, and deep learning for optimizing resource allocation could further improve the EE and performance of mobile networks. As a result, NOMA emerges as a promising solution for 5G and especially future 6G networks, with the potential to significantly reduce network EC and optimize resource utilization.

4.8 Application of Network Slicing Technology for Improving Energy Efficiency in Mobile Networks

As presented in Chapter 3, Section 3.8, network slicing (NS) is a technology that allows a single physical mobile network to be logically divided into multiple virtual networks (slices). Each network slice is customized to meet the performance requirements of specific use cases such as enhanced mobile broadband (eMBB), ultra-reliable low-latency communication (URLLC), massive machine-type communications (mMTC), emergency communications, or smart cities, industry, health, agriculture, etc. Figure 4.9 presents scaling of mobile network resources and activity state of network elements based on the specific network slice needs. NS allows on-demand allocation of network resources, which ensures that only the necessary computing and radio resources are activated and utilized within the active network slice. Unused network resources in one network slice can be switched off or reallocated to other network slices, which contributes to reducing unnecessary mobile network EC.

NS can also enable energy-efficient network traffic management, since traditional unsliced networks handle all types of traffic within a single common infrastructure and this leads to inefficient network energy use. NS isolates traffic types within a corresponding network slice and this isolation ensures that each network slice operates only with the required level of

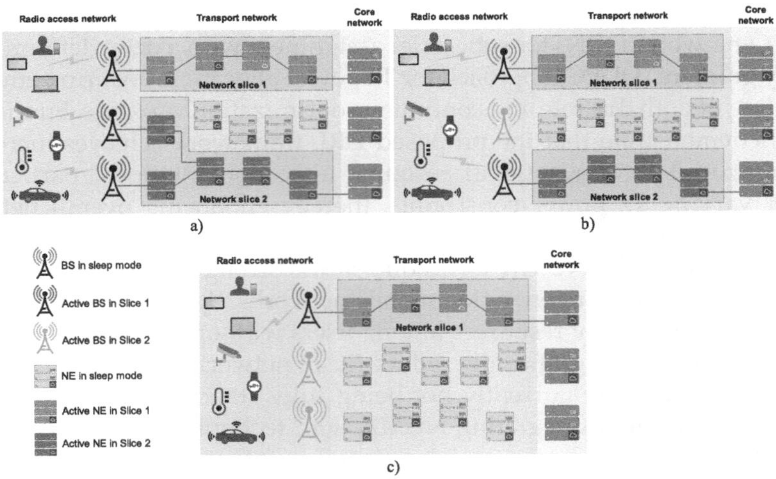

FIGURE 4.9
Scaling of mobile network resources and activity state of network elements based on the specific network slice needs, (a) active state of two network slices exploiting full network resources, (b) active state of two network slices, (c) active state of one network slice [107]

processing power and bandwidth (Figure 4.9). For example, data-intensive services (such as virtual reality [VR], extended reality [XR], or video streaming) can be isolated in high-power demanding network slices, while low-power demanding network slices can isolate services such as those related to IoT or sensor network slices that consume minimal energy. Also, such isolation of specific network traffic in dedicated network slices can contribute to reducing network congestion, which can lead to lowering the BS and UD Tx power requirements, and consequently to improving network EE.

Additionally, NS is very suitable for implementation with AI and ML techniques, which can be used for predicting traffic loads and automatically adjusting network slice resources in real time. This can ensure that power-demanding network slices operate only when necessary (Figure 4.9), thus optimizing the overall operational time of network elements of such network slices. Such AI-controlled network energy management can prevent excessive mobile network EC through predictive resource allocation which exploits dynamic activation and deactivation of network slices.

Thus, NS technology represents a promising concept for improving EE in 5G and future 6G networks, enabling flexible and dynamic resource allocation by creating virtual network segments tailored to specific UD or application needs. This enables optimized use of network capacity and can contribute to reducing excessive network EC and improving performance in high-traffic network conditions. However, the massive implementation of NS still has not taken place in practice, since different challenges related to the implementation of NS in mobile networks need to be resolved. Among the most demanding challenges is solving the network slice management complexity that needs to balance network EE and satisfying specific network slice QoS requirements. Also, NS imposes high initial CAPEX that is reflected in the need for ensuring network virtualization, cloud-native infrastructure, and AI-based orchestration. Additionally, security risks related to the need to have strong cybersecurity of network slices demand security innovations that will enable full isolation among network slices. In this context, different research results have been presented to solve the main challenges related to the implementation of NS for improving network EE.

In [108], a framework for energy-efficient and secure NS is proposed. The framework is based on utilizing an integer linear programming (ILP) model to minimize network EC while meeting strict security and QoS requirements. The framework enables mobile network resource optimization by effectively embedding virtual network functions (VNFs) within network slices, thus allowing mobile operators to adapt network resources to the specific UD needs of each network slice. Simulation results presented in [108] show that the proposed approach achieves significant energy savings compared to traditional network management methods, which contributes to improving overall network EE.

Furthermore, it is shown in [109] that NS technology can have a crucial role in supporting IoT applications in smart cities by addressing diverse performance requirements. In [109], the authors propose an automated NS system driven by ML to dynamically adapt network resources to the specific needs of IoT devices. This system utilizes network slice state prediction, anomaly detection, and slicing scaling to optimize mobile network resource allocation and improve EE. By intelligently managing network slices, the framework supports a wide range of IoT applications, while significantly reducing mobile network EC and enhancing overall network performance.

In [110], it is shown that implementation of NS technology is particularly relevant in the context of the Industrial Internet of Things (IIoT), since NS can offer customized mobile network resource management for diverse applications such as smart factories, smart energy systems, and smart transportation. In [110], the authors highlight the use of intelligent NS to manage these industrial applications, using AI-driven techniques for mobile network resource allocation and orchestration. The proposed framework enables dynamic adaptation of mobile network segments to meet application-specific requirements, thus improving mobile network EE and reducing EC across all IIoT services. These improvements ensure that industrial systems can effectively deal with the demands of complex and changing industrial environments, while maintaining optimal performance and sustainability.

The EE optimization through the implementation of NS within a C-RAN architecture is discussed in [111]. The authors propose an end-to-end framework for energy-efficient NS by jointly allocating communication and computation resources across all network functions, which span from radio access to centralized cloud processing parts of the mobile network. The framework considers constraints such as delay requirements, fronthaul and midhaul transmission capacity, and processing capacity to balance EC across network segments. Simulation results show that such an approach reduces mobile network EC while meeting QoS constraints. This study highlights the importance of NS in improving resource utilization and EE for 5G and future mobile network architectures.

Additionally, fog computing as one of the advanced concepts in mobile networks, based on moving computing resources closer to users, can also use NS technology to optimize mobile network energy efficiency. In [112], the authors propose dynamic NS within fog computing systems, coordinated by a regional orchestrator to manage workload distribution among fog nodes. The proposed system also leverages energy harvested from the environment to provide computational services to local UDs, which contributes to the further reduction of network EC. By employing an algorithm exploiting a belief-state partially observable Markov decision process, the model proposed in [112] enables efficient mobile network resource allocation under dynamic conditions. Simulations using data from real BSs demonstrate significant improvements in the workload processing capacity of mobile networks.

While fog computing demonstrates the potential for joint implementation with NS to improve decentralized resource management and EE in specific local mobile network segments, the challenge of constant and large increases of 5G network EC requires the implementation of comprehensive energy-aware strategies that will span a complete mobile network. The study in [113] presents a novel adaptive energy management framework for 5G NS, focusing on optimizing network power consumption while incorporating renewable energy sources in network power supply. An architecture based on NFV and SDN introduced in [113] enables dynamic allocation of energy resources across different network slices. The study also proposes several energy management strategies, including service quality degradation, service rejection, intra-slice prioritization, and time-shifted execution, to adapt network EC to available power. Furthermore, an ILP model is formulated to optimize energy distribution among network slices from renewable and non-renewable energy sources. The simulation results confirm that the proposed NS model significantly reduces EC in mobile networks, particularly in multi-tenant environments where energy-efficient NS is critical for ensuring OPEX reductions.

In conclusion, NS technology can be an important enabler for achieving improved EE of future mobile networks. NS can contribute to the improvement of mobile network EE by enabling dynamic network resource allocation, intelligent traffic management, and AI-driven network resource optimization. Through dynamic resource allocation, adaptation to different services and applications, and the use of advanced energy management methods, NS enables energy optimization in scenarios with high traffic and complex user demands. The studies presented in this section demonstrate that NS can be used in various environments and use cases, including smart cities, IoT, and IIoT systems, with the application of advanced algorithms and technologies such as ML and environmental energy harvesting. Thus, the NS not only can contribute to reducing mobile network EC but also can provide flexible and reliable realization of next-generation networks, which is essential for the development of 5G and future 6G networks.

4.9 Impact of Space-Satellite Constellations on Mobile Network Energy Consumption

The deployment of space-based 5G network constellations mainly consists of a large number of low Earth orbit (LEO) satellites that orbit between 480 km and 2,000 km above the Earth. Figure 4.10 presents the implementation of space-satellite constellations as a complement to terrestrial mobile networks. The ground segment of the 5G satellite communication network consists of terrestrial (ground) stations that are allocated worldwide.

The terrestrial relay (ground) stations enable connection to space network segments over ground-based satellite antennas pointing directly at the satellites for wireless signal transmission and interconnection with the ground core network infrastructure that is established over high-speed fiber-optic links (Figure 4.10). These LEO satellites harvest renewable energy through a solar power supply and orbit around Earth making at least 11.25 orbits around the Earth per day. This results in non-constant LEO satellite contact with a ground transmitter, and for maintaining seamless coverage,

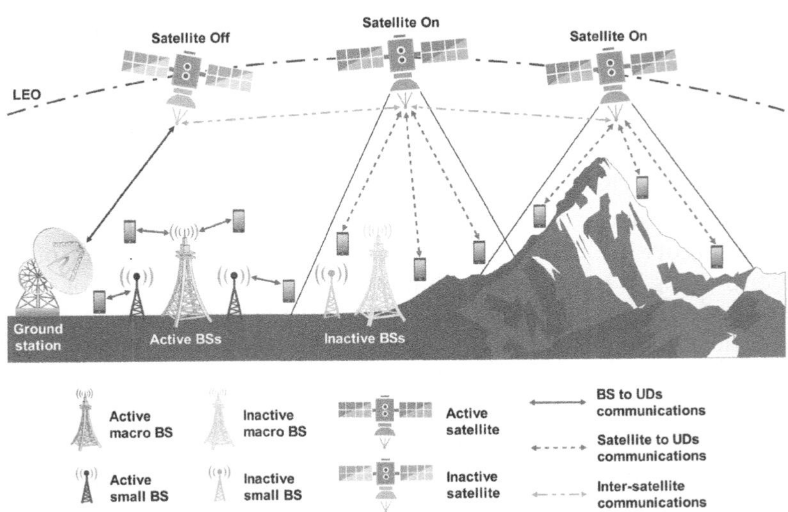

FIGURE 4.10
Implementation of space-satellite constellation as a complement to terrestrial mobile networks

communication from the ground is transmitted among LEO satellites across the satellite constellation.

The 5G network satellite Internet brings many advantages in mobile network coverage and capacity on the global level. These advantages are seen in 5G satellite networks complementing terrestrial 5G network infrastructure in extending network capacity in urban areas or coverage to remote areas such as oceans, hardly accessible regions, or regions uncovered with mobile network signal due to lack of commercial interest of mobile operators to invest in terrestrial network infrastructure in such regions (Figure 4.10). For example, in 2020, 63% of rural households on the global level lacked Internet access due to the infrastructure challenges related to the inability to connect households with underground cables or by some type of mobile network [114]. Thus, 5G network satellite Internet, while improving traditional terrestrial networks' resilience and scalability, can provide unprecedented connectivity and support for the realization of future advanced applications.

The offering of commercial 5G network satellite Internet started in 2020, when aerospace company SpaceX, through the Starlink satellite constellation composed of 1,000 LEO satellites, began offering 5G satellite Internet to customers. This rolled out a new era of the 5G network satellite Internet race, which was followed by the Amazon company announcement on launching the Kuiper satellite constellation composed of 3,236 LEO satellites [115] and the OneWeb company announcement on launching a satellite constellation composed of 648 LEO satellites [116]. Undoubtedly, the 5G network satellite Internet race will continue in the future and more companies, governments, organizations, or countries will be involved in developing and offering satellite Internet based on 5G and Beyond 5G (B5G) mobile network technologies. The massive global rollout of the 5G and B5G network satellite Internet in the future, besides jeopardizing terrestrial mobile telecom operator revenues, is expected to have an impact on terrestrial mobile network EE.

Satellite-based mobile networks can contribute to the improvement of terrestrial-based mobile networks EE in multiple ways. Firstly, traditional mobile networks require energy-intensive BSs to cover remote or hardly accessible areas. The satellite-based mobile network constellations can be used for covering remote areas lacking terrestrial mobile network signal coverage (Figure 4.10). This minimizes the need for deployment of a large number of BSSs in remote areas that often lack grid power supply and rely on extremely energy-inefficient off-grid diesel generator power supply.

Secondly, satellite-based mobile networks can contribute to traffic offloading of congested terrestrial mobile network parts, which are mostly characteristic for urban and dense urban mobile network areas (Figure 4.10). This offloading can contribute to reducing terrestrial mobile network EC, since terrestrial BSs can enter sleep (low-power) states when satellite signal coverage and capacity is available to UDs in terrestrial mobile networks. To accomplish such EE mobile network resource scheduling, network signaling exchange and resource management need to be scheduled on the level of the terrestrial and space-satellite segments of the mobile network. This opens a completely new field of research that can be based on the development of AI-driven algorithms, capable of dynamically allocating UD connections among space-satellite and terrestrial BSs and, thus, enabling optimized energy usage of terrestrial mobile network segments. Also, the implementation of satellite-based mobile network constellations enables the less dense deployment of terrestrial network infrastructure (such as BSSs), which also contributes to reducing the overall EC of mobile networks, especially during peak traffic periods.

Thirdly, besides direct UD-to-satellite connectivity, satellite-based mobile networks can enable backhaul connectivity alternatives to UDs served by private 5G and B5G networks or other wireless network types such as wireless local area network (WLAN), long range (LoRa), narrowband internet of things (NB-IoT), etc. (Figure 4.10). Thus, satellite-based mobile networks can serve as an energy-efficient backhaul alternative by replacing high-energy microwave or fiber-optic backhaul links. This contributes to reducing mobile network EC in the segment of long-range terrestrial microwave backhaul transmissions and in the segment of fiber-based long-haul connections. Also, direct UD-to-satellite connectivity eliminates the need for energy-consuming relays and gateways needed in terrestrial IoT networks. This can be particularly useful in specific use case scenarios of IoT networks, where a large number of individual sensing devices can exchange data over satellite-based mobile networks.

However, the satellite-based mobile networks can also impose a negative impact on the UDs' EC. This is reflected in direct UD-to-satellite communication, in which UD signals must be transmitted at higher Tx power levels to reach satellite antennas which are far more distant than antennas of terrestrial BSs (Figure 4.10). This contributes to the faster discharging of UD batteries and the shortening of the UD battery lifetime. Part of the solution can be in the satellite wireless signal Tx power control, which can dynamically adjust the level of Tx power and signal beam direction through satellite beamforming toward a specific UD or group of UDs. Also, satellite power consumption will be significantly impacted by the number of simultaneously served UDs. An increase in the number of served UDs will contribute to an increase in satellite EC, and optimal scheduling of satellite power supply according to a load of satellite BS(s) will be one of the important research challenges in the future. The development of advanced dynamic and adaptive Tx power control algorithms will be essential to optimize the EE of satellites. These algorithms will need to enable satellites and their corresponding BS(s) to enter sleep mode when no UDs are served and to enter active mode of operation when UDs are detected within the satellite's coverage area.

Additionally, satellite-based mobile networks will not be appropriate for use cases and applications whose operation demands extremely low signal latencies. For example, the safety of autonomous vehicles can be impacted by even a slight delay in communication between UD and satellite. Such applications based on near-instantaneous decision-making cannot rely on satellite-based mobile networks, which is the reason why, in the future, satellite-based mobile networks will remain a complement and not a full replacement of terrestrial mobile networks.

Above all, the exploitation of satellite-based mobile network constellations for complementing terrestrial mobile networks is energy efficient by itself, since LEO satellites as the main elements of mobile network space-satellite constellations are power supplied by sun energy as one of the renewable energy sources (Figure 4.10). Thus, in the operating phase, the space segment of mobile network satellite constellations is highly energy efficient, especially when compared with the EE of the terrestrial-based mobile network segment during the mobile network operation phase. However, it is worth emphasizing that the energy needed for producing hundreds or thousands of satellites and sending them into the low Earth orbit is significant and cannot be neglected. However, it is reasonable to assume that this vast energy consumption will be annulled during the lifetime of the satellite operation powered by renewable energy sources. In addition, the power consumption of terrestrial (ground) stations makes a ground segment of the satellite network a contributor to the overall mobile network EC. Since the number of satellite network ground stations is significantly lower than the number of terrestrial BSSs, it is reasonable to assume that the contribution of ground stations' EC to the overall mobile network EC will be negligible in comparison with the EC contribution of BSSs.

Satellite-based mobile network constellations therefore have the potential to significantly enhance terrestrial mobile network EE. This enhancement can be achieved through reduced usage of high energy-consuming terrestrial network infrastructure, satellite backhauling, and optimized satellite-terrestrial traffic management. To achieve these enhancements, different AI-based energy optimization approaches need to be developed to support sustainable and energy-efficient mobile communications in both terrestrial and space segments of the mobile networks.

4.10 Application of Reconfigurable Intelligent Surfaces for Improving Energy Efficiency of Mobile Networks

The rising demands on 5G and beyond (B5G/6G) networks require innovative solutions to enhance EE, while simultaneously satisfying requirements for increasing data rates and capacity. One of the emerging technological approaches addressing these challenges is the use of reconfigurable intelligent surface (RIS), also known as intelligent reflecting surface (IRS) [117] or programmable metasurface [118]. RIS consists of planar arrays comprising

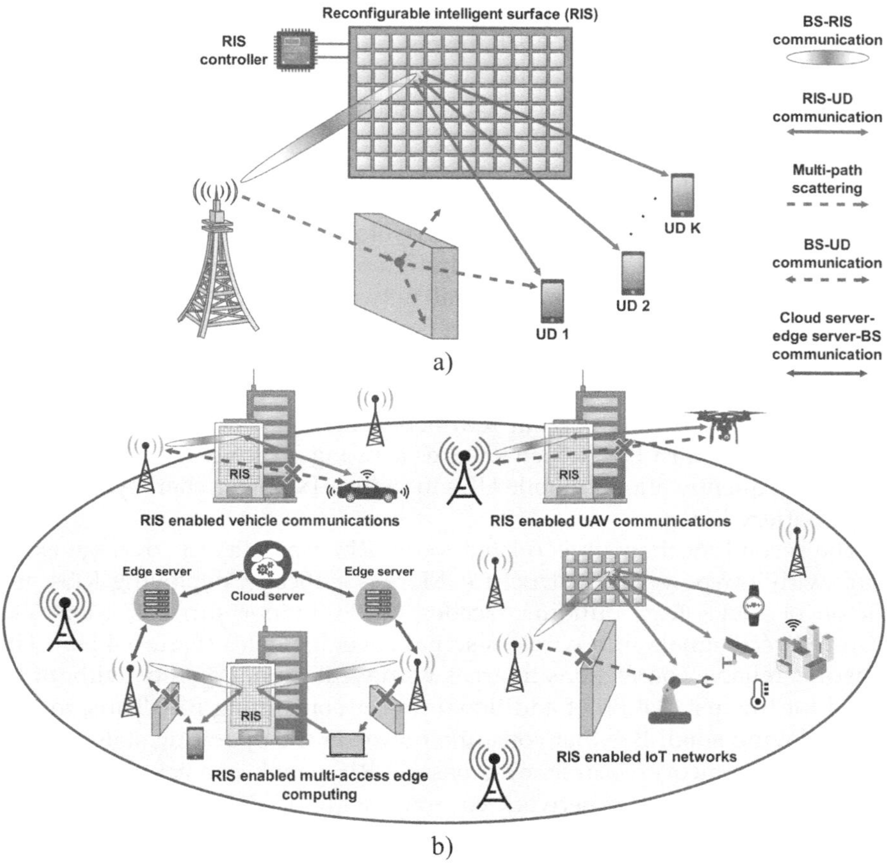

FIGURE 4.11
Implementation of RIS technology in the wireless network for (a) basic operating principle with RIS controller, (b) different use cases

numerous reflective elements, each connected to a tunable chip capable of dynamically controlling their reflection characteristics. Figure 4.11 visualizes an implementation of the RIS technology in wireless networks for basic operating principles and different use cases.

As illustrated in Figure 4.11, RISs are typically composed of metamaterial elements that can be reconfigured electronically to reflect, refract, and scatter radio signals in a desired manner. Based on an RIS controller that enables flexibly manipulating with signal characteristics such as phase, amplitude, and polarization (Figure 4.11a), the RIS technology facilitates active shaping of the wireless propagation environment [118]. RIS units operate predominantly as nearly passive devices, consuming minimal power primarily for the control circuitry of tunable chips controlled with the RIS controller. Thus, the RIS technology has the potential to significantly enhance EE in mobile networks through different mechanisms.

The first mechanism is related to the concept of RIS technology that in essence contains RIS elements which intelligently adjust the amplitude, phase, or polarization shifts of reflected signals, creating a constructive wireless signal superposition at the receiver. This enables a higher signal quality at the position of the wireless signal recipient for the lower Tx power of the transmission device, which can have a twofold contribution to the improvement of network EE. The first contribution related to RIS implementation is in the possibility of minimizing overall network EC by reducing the need for BS transmission at the higher Tx powers, which, in traditional networks lacking RIS implementation, increases BS Tx power to compensate for weak signal quality and thus contributes to an increase in network EC. Additionally, RIS technology through signal redirects contributes to enhancing signals transmitted toward BSs, which ensures a stronger signal of uplink channel and consequently allows mobile UDs to reduce Tx power, thereby extending their battery life.

The second mechanism is related to the RIS capability of coverage extension with low-power infrastructure. More specifically, deploying RISs near or on obstacles (e.g., building facades, walls, indoor surfaces, roofs, etc.) can redirect signals into wireless signal coverage holes (Figure 4.11b). This ensures reliable connections in areas with weak signal quality, without the need for the installation of additional power-consuming BSs. Thus, instead of deploying small BSs that consume power in the operating state to cover some signal uncovered area, near-passive RIS panels can act as low-power reflectors and improve network coverage without allocation of additional energy-demanding network infrastructure.

The third mechanism that contributes to the improvement of mobile network EE through the implementation of RIS is related to reducing network computational burden by decreasing signal processing overhead at BSs. It is already stated that BSs on average can operate at lower Tx power, since the implementation of RIS can enhance signal strength and reduce fading at the location of UDs. This contributes to reducing the complexity of mobile network interference management, which consequently reduces operating complexity at the BS level and can further contribute to reductions of BS power consumption.

The fourth mechanism related to the RIS capability of improving network EE is related to the possibility that RIS can contribute to the minimization of transmission errors and the number of retransmissions in mobile networks. This is due to the RIS capability to improve wireless signal quality at the location of UDs (Figure 4.11). More specifically, poor signal conditions can often lead to packet losses and retransmissions, which are processes that contribute to consuming additional processing power of BSs and consequently impact on increase of mobile network EC. Since RIS enhances the SNR at the receiver side, reducing errors and data retransmissions leads to lower BS EC. Also, the implementation of RIS can contribute to the reduction of EC in edge computing (Figure 4.11b). Since RIS reduces retransmissions and enhances the reliability of communication links for multi-access edge computing (MEC), this leads to lower energy consumption in cloud-edge interactions, making energy-hungry AI/ML applications more sustainable.

The fifth mechanism for improving mobile network EE through the implementation of RIS is related to the capability of RIS to contribute to dynamic load balancing and smart network resource allocation. More specifically, RIS can be implemented to dynamically adapt to real-time network conditions and this adaptation can contribute to optimizing wireless uplink signal directing and in reducing network energy waste. By intelligently steering signals to less congested network BSs (Figure 4.11), RIS can help in the optimal balancing of network traffic and such an approach further contributes to reducing the need for allocation of power-demanding BSs in overloaded areas. In addition, RIS can contribute to the reduced need for complex antenna arrays on the BS or UD side. More precisely, on the BS side, instead of using power-hungry multi-antenna MIMO or mMIMO arrays, BSs can rely on RIS-assisted reflections to achieve transmission toward UDs with higher data rates over fewer active antennas. This can have a significant impact on the improvement of BS energy efficiency, due to reducing the power required for mMIMO signal processing. On the UD side, instead of relying on complex and energy-demanding MIMO processing, UDs can leverage intelligent reflections of dedicated signal beams within RIS-assisted networks, in order to explore energy-aware beamforming. This enables UDs to communicate efficiently with reduced computational overhead over fewer antennas, enabling data transmission with lower EC that contributes to extending the UD battery life.

The sixth mechanism that can contribute to improving network EE through the implementation of RIS is related to enhancing wireless charging/energy harvesting and relaying techniques. Implementation of RIS can optimize the focusing of electromagnetic waves for wireless power transfer (WPT) and SWIPT. By directing RF signal energy efficiently to charging UDs by support of RIS, energy-efficient wireless charging solutions can be enabled and this can contribute to reducing battery drain in UDs. Also, the implementation of RIS can contribute to the performance improvements of SWIPT, in terms of more precise transfer of information, which can contribute to the overall improvement of network EC.

Additionally, RIS can enable the wireless signal relaying concept that is based on the implementation of wireless signal repeaters. Unlike traditional active wireless signal repeaters or relays, RIS operates almost passively, which means that RIS does not generate additional signals through a high-power demanding amplification process, but instead reflects and redirects existing signals (Figure 4.11). Thus, RIS has the potential to significantly reduce overall mobile network EC in comparison with traditional relay systems whose operating principle is based on power-intensive wireless signal amplification [119]. This low operational power makes RIS particularly suitable for improving EE in wireless communication networks.

The implementation of RIS can be an important technology for the development of more energy-efficient mobile networks and UDs. As 5G-Advanced and 6G networks continue to evolve, RIS will play a crucial role in creating sustainable and energy-efficient communication systems. However, more comprehensive investigations and further research in the future need to be done to explore all possibilities of RIS technology in improving mobile network EE. Some of the initial research attempts are reviewed in the following subsection.

4.10.1 Research Overview on the Application of Reconfigurable Intelligent Surfaces for Improving Energy Efficiency of Mobile Networks

Besides almost passive RIS, there is the possibility of implementing active RIS, which can further amplify the impinging RF waves by integrating signal amplifiers into the operation of RIS unit cells [119]. Although active RIS can overcome signal path loss more effectively, it implies higher implementation costs and increased EC, which may reduce some of the previously described mechanisms that can contribute to improving mobile network EE.

The integration of RIS into cellular networks offers several key benefits, including improved signal coverage, enhanced SE, reduced infrastructure deployment costs, and improved EE [120, 121]. By strategically deploying RIS, mobile network operators can effectively overcome wireless signal coverage gaps and obstacles (Figure 4.11), especially in dense urban environments or indoor scenarios where conventional network solutions may be either impractical, expensive, or energy inefficient.

In addition to these general advantages of RIS technology, when adequately optimized, RIS technology has shown in previously published research a significant potential for improving mobile network EE. Numerous studies have been performed on changing uncontrolled wireless environments into dynamically controllable communication channels through the implementation of RIS, thereby significantly improving communication reliability and wireless channel capacity (Table 4.5). Research presented in [121] has been conducted using different methods to increase mobile network EE in RIS-assisted networks, with results highlighting the large potential of this technology. Similarly, the authors in [122] demonstrated considerable EE improvements achievable through RIS technology. Specifically, RIS-based systems employing joint optimization strategies for Tx power and RIS element configurations have shown significant EE enhancements. For instance, numerical results of this research showed that RIS-based communication can provide up to 300% higher EE compared to conventional relay-assisted systems. These findings highlight the great potential of RIS technology as a sustainable and energy-efficient approach to future wireless network designs.

TABLE 4.5

Summary of Research on Mobile Network EE Optimization in RIS-Assisted Networks

Reference	Focus Area	Key Contributions
[122]	RIS-based optimization of network EE	Demonstrated that RIS-based systems with joint optimization of Tx power and RIS configurations can achieve up to 300% higher EE than relay-assisted systems
[123]	Resource allocation in distributed RIS	Developed two network EE maximization algorithms for distributed RIS applications, optimizing RIS phase and activation patterns to enhance network EE in limited power and high-density user scenarios
[124]	On-off RIS power dissipation model	Proposed a novel RIS power dissipation model accounting for on-off RIS state differences, leading to more accurate EE optimization by minimizing BS Tx power and RIS power dissipation
[125]	Distributed RIS framework (DRF) for indoor THz networks	Utilized 3D ray-tracing to optimize RIS placement in THz networks for improving SNR and QoS, while also significantly enhancing EE for IoT applications
[126]	RIS integration in distributed MIMO (D-MIMO) systems	Investigated implementation of RIS in D-MIMO systems demonstrated that optimized RIS placement and user-centric access point (AP) clustering can improve EE and coverage, reducing the need for additional active APs
[127]	Hierarchical reinforcement learning (HRL)-based RIS optimization for 5G RAN	Developed an HRL framework combining RIS and sleep control techniques to optimize BS operations, which resulted in more than doubling network EE and throughput improvements in high-traffic scenarios
[128]	RIS-based hybrid precoding for terahertz (THz) communications	Introduced a deep learning-based multiple discrete classification (DL-MDC) hybrid precoding method for enhancing network EE in RIS-assisted THz systems, utilizing a parallel DNN system to efficiently solve sum-rate maximization while reducing computational complexity
[129]	RIS implementation in non-terrestrial networks (NTNs) (including LEO satellites, UAVs)	Reviewed RIS integration in NTNs emphasized passive signal redirection to reduce EC, dynamic on-off control for adaptive power management, and ML-based optimizations for improving network EE

Moreover, recent research in [123] has explored the network resource allocation problem in wireless networks by employing distributed RIS configurations, with multiple RIS units spatially distributed to optimize network EE (Figure 4.11b). The optimization involves dynamically controlling the on-off status of each RIS and adjusting their reflection coefficients to maximize system EE with minimal user data rate restrictions (Table 4.5). Two computationally efficient network EE maximization algorithms have been proposed. The first one is designed for the single-user case and the other one is developed for the multi-user case. Both algorithms consider RIS phase optimization and optimal RIS activation patterns. Numerical results obtained by these studies have shown that such jointly optimized distributed RIS strategies significantly outperform in terms of network EE, conventional networks lacking implementation of RIS. This is especially confirmed in scenarios involving limited Tx power and increased user density, which highlights their potential to improve EE in future wireless networks.

While prior studies have demonstrated the benefits of optimizing distributed RIS configurations for improving network EE, most existing works have overlooked the differences in power consumption between the on (active) and off (inactive) states of RIS physical elements. This results in not accurately presenting the characteristics of these elements in the real world. To address this limitation, recent work in [124] has introduced a novel on-off aware power dissipation model for RIS elements, which accounts for the differences in power consumption between the on and off states of these elements (Table 4.5). This novel RIS power model allows for more realistic EE optimization by jointly minimizing the active Tx power of the BS and passive RIS power dissipation. Simulation results from this study validated the effectiveness of the proposed model and associated algorithms, demonstrating improved performance in terms of network EE, compared to traditional optimization approaches that do not take these practical power consumption considerations into account.

While refinement of power dissipation models increases the network EE, further advances in RIS implementation strategies may enable even greater gains regarding network EC reductions. One promising direction involves optimizing RIS configurations for specific environments, such as high-frequency THz communications, where unique wireless communication challenges and opportunities arise. Hence, a recent study in [125] has proposed a distributed RIS framework (DRF) for indoor THz wireless communication systems (Table 4.5). Utilizing 3D ray tracing to simulate realistic propagation environments, including dynamic human obstructions, the proposed DRF optimizes RIS configurations with the aim to improve network EE and ensure adequate QoS. Numerical analyses demonstrated significant improvements in SNR and overall system performance when compared with conventional "natural" wireless signal reflections. Such distributed RIS configurations promise substantial gains in network EE that are particularly suitable for high-frequency THz communications in future IoT applications, thus highlighting the broader potential of RIS technologies in different practical use case scenarios (Figure 4.11b).

The authors in [126] have further examined the integration of RIS into distributed MIMO (D-MIMO) systems, specifically focusing on indoor environments with high obstacle density and large user populations (Table 4.5). Utilizing dynamic, user-centric wireless AP clustering and optimized RIS placement techniques, this approach significantly enhances network EE and scalability of indoor wireless networks. Numerical analyses confirmed that even with a relatively small number of RIS elements, substantial improvements in average network EE and service coverage are achievable, making RIS integration an attractive, energy-efficient alternative to network implementations based on increasing the number of installed active APs. Such results emphasize the practicality and benefits of RIS-assisted D-MIMO deployments in achieving sustainable and high-performance indoor wireless networks.

While conventional optimization techniques have demonstrated significant EE gains of networks exploiting RISs, recent research has explored the use of ML to further enhance system adaptability and performance. According to the research presented in [127], recent advances have introduced hierarchical reinforcement learning (HRL) to optimize the management of RIS-assisted heterogeneous 5G RAN (Table 4.5). The HRL framework strategically combines RIS with BS sleep control techniques, employing a meta-controller to determine the on-off status of small BSs and a sub-controller to dynamically adjust small BS Tx power levels. Simulation results indicate that this HRL-based RIS-assisted scheme significantly reduces network EC, while more than doubling the network EE compared to traditional sleep control methods lacking implementation of RISs. This research demonstrates that integrating RIS technology with advanced AI-based methods could be an effective approach to improve network EE in future mobile network architectures.

In another study [128], the authors proposed a RIS-based hybrid precoding architecture for THz communications to enhance network EE by replacing power-hungry phased arrays with an energy-efficient RIS for analog beamforming (Table 4.5). Given the non-convex nature of optimizing discrete phase shifts, the study reformulated the sum-rate maximization problem into several distinct classification tasks and introduced a deep learning-based multiple discrete classification (DL-MDC) hybrid precoding method to efficiently solve them. The proposed approach employs a parallel DNN system to facilitate the optimization process, significantly reducing computational complexity while maintaining performance. This integration of RIS with ML techniques offers a promising direction for developing adaptable and energy-efficient THz communication, supporting the evolution of sustainable 6G networks.

While previous research has mostly focused on RIS applications in terrestrial networks, new studies highlight its potential in non-terrestrial environments. The extension of RIS technology to non-terrestrial networks (NTNs) presents new opportunities to improve network EE, especially in scenarios where conventional network infrastructure is impractical for implementation or demands high EC. Recent advances in RIS technology have shown significant potential for increasing mobile networks EE, especially when integrated with NTNs. As reviewed in [129], studies have explored the integration of RISs in high-altitude platforms (HAPs) including LEO satellites and unmanned aerial vehicles (UAVs). The research presented in [129] highlights that the implementation of RISs with NTNs has the potential to significantly reduce network EC through passive signal redirection, thereby eliminating the need for energy-intensive active transmission (Table 4.5). This passive beamforming capability enables network coverage expansion with minimal power consumption of network elements, making RIS-assisted NTNs particularly effective for remote and inaccessible areas where traditional infrastructure deployment is impractical.

The survey in [129] also reviews studies that highlight the importance of dynamic on-off control of RIS elements activity (Table 4.5). Such an approach enables adaptive network power consumption management based on traffic demands, which further contributes to improving the EE of the entire system. Additionally, [129] discusses the research on ML-based optimizations in RIS-assisted NTNs, which through facilitating intelligent adaptation to environmental conditions maximizes network EE, while maintaining network connectivity and reliability. These findings suggest that RIS technology has the potential to significantly contribute to the development of sustainable NTNs, and that implementation of RIS technology is aligned with broader goals related to increasing EE of future 6G communications.

Therefore, the RIS technology has shown significant potential for improving mobile network EE through intelligently shaping the wireless signal propagation environment. By minimizing reliance on power-hungry mobile network components and by optimizing signal transmission and integrating it with AI-based power management strategies, RIS can contribute to more sustainable and energy-efficient wireless communication systems. To fully realize these benefits, future research should focus on advanced ML-based optimization techniques and practical implementation strategies that maximize mobile network EE, while maintaining network performance and scalability.

References

1. I. P. Chochliouros *et al.*, "Energy efficiency concerns and trends in future 5G network infrastructures," *Energies (Basel)*, vol. 14, no. 17, 2021. doi: 10.3390/en14175392.
2. GSMA Intelligence, "Going green: Measuring the energy efficiency of mobile networks," 2024. Accessed: Aug. 18, 2024. [Online]. Available: https://data.gsmaintelligence.com/api-web/v2/research-file-download?id=79791160&file=270224-Measuring-energy-efficiency-of-mobile-networks.pdf

3. NGMN Alliance, "Network energy efficiency phase 2," 2023. Accessed: Aug. 18, 2024. [Online]. Available: https://www.ngmn.org/wp-content/uploads/NGMN_Network_Engergy_Efficiency_Phase2.pdf
4. J. Wu, Y. Zhang, M. Zukerman, and E. K. N. Yung, "Energy-efficient base-stations sleep-mode techniques in green cellular networks: A survey," *IEEE Communications Surveys and Tutorials*, vol. 17, no. 2, 2015. doi: 10.1109/COMST.2015.2403395.
5. J. O. Ogbebor, A. L. Imoize, and A. A. A. Atayero, "Energy efficient design techniques in next-generation wireless communication networks: Emerging trends and future directions," 2020. doi: 10.1155/2020/7235362.
6. M. H. Alsharif, J. Kim, and J. H. Kim, "Green and sustainable cellular base stations: An overview and future research directions," 2017. doi: 10.3390/en10050587.
7. M. Isaksson, *Radio Frequency Power Amplifiers Behavioral Modeling, Parameter-Reduction, and Digital Predistortion*. Stockholm: KTH Royal Institute of Technology in Stockholm, 2007.
8. Q. F. Cheng, S. K. Zhu, and H. F. Wu, "Investigating the global trend of RF power amplifiers with the arrival of 5G," in *2015 IEEE International Wireless Symposium, IWS 2015*, 2015. doi: 10.1109/IEEE-IWS.2015.7164624.
9. A. Vasjanov and V. Barzdenas, "A review of advanced CMOS RF power amplifier architecture trends for low power 5G wireless networks," 2018. doi: 10.3390/electronics7110271.
10. P. Y. Chen *et al.*, "Learning to Compensate: A deep neural network framework for 5G power amplifier compensation," in *IEEE International Conference on Communications*, 2021. doi: 10.1109/ICC42927.2021.9500277.
11. V. Chamola, S. Patra, N. Kumar, and M. Guizani, "FPGA for 5G: Re-configurable hardware for next generation communication," *IEEE Wireless Communications*, vol. 27, no. 3, 2020. doi: 10.1109/MWC.001.1900359.
12. P. Visconti, R. Velazquez, C. Del-Valle-Soto, and R. De Fazio, "FPGA based technical solutions for high throughput data processing and encryption for 5G communication: A review," *Telkomnika (Telecommunication Computing Electronics and Control)*, vol. 19, no. 4, 2021. doi: 10.12928/TELKOMNIKA.v19i4.18400.
13. A. Martinen, K. Ruttik, and R. Jantti, "Saving energy in base station with non real time operation system in Cloud-RAN," in *2014 IEEE Globecom Workshops, GC Wkshps 2014*, 2014. doi: 10.1109/GLOCOMW.2014.7063584.
14. N. Amani, H. Pedram, H. Taheri, and S. Parsaeefard, "Energy-efficient resource allocation in heterogeneous cloud radio access networks via BBU offloading," *IEEE Transactions on Vehicular Technology*, vol. 68, no. 2, 2019. doi: 10.1109/TVT.2018.2882466.
15. J. Gao, X. Guan, S. Zhang, and X. Meng, "Resource allocation optimization based on energy efficiency in green cloud radio access network," *Wireless Communications and Mobile Computing*, vol. 2022, 2022. doi: 10.1155/2022/8932961.
16. R. Guerra-Gomez, S. Ruiz-Boque, M. Garcia-Lozano, and J. O. Bonafe, "Machine learning adaptive computational capacity prediction for dynamic resource management in C-RAN," *IEEE Access*, vol. 8, 2020. doi: 10.1109/ACCESS.2020.2994258.
17. J. Lorincz, A. Capone, and D. Begušić, "Heuristic algorithms for optimization of energy consumption in wireless access networks," *KSII Transactions on Internet and Information Systems*, vol. 5, no. 4, 2011. doi: 10.3837/tiis.2011.04.001.

18. R. Wang, J. S. Thompson, H. Haas, and P. M. Grant, "Sleep mode design for green base stations," *IET Communications*, vol. 5, no. 18, 2011. doi: 10.1049/iet-com.2011.0104.
19. J. Lorincz, A. Capone, and D. Begusic, "Impact of service rates and base station switching granularity on energy consumption of cellular networks," *EURASIP Journal on Wireless Communications and Networking*, vol. 2012, no. 1, 2012. doi: 10.1186/1687-1499-2012-342.
20. K. Son, H. Kim, Y. Yi, and B. Krishnamachari, "Base station operation and user association mechanisms for energy-delay tradeoffs in green cellular networks," *IEEE Journal on Selected Areas in Communications*, vol. 29, no. 8, 2011. doi: 10.1109/JSAC.2011.110903.
21. S. Bhaumik, G. Narlikar, S. Chattopadhyay, and S. Kanugovi, "Breathe to stay cool: Adjusting cell sizes to reduce energy consumption," in *Proceedings of the 1st ACM SIGCOMM Workshop on Green Networking, Green Networking '10, Co-located with SIGCOMM 2010*, 2010. doi: 10.1145/1851290.1851300.
22. M. F. Hossain, K. S. Munasinghe, and A. Jamalipour, "Toward self-organizing sectorization of LTE eNBs for energy efficient network operation under QoS constraints," in *IEEE Wireless Communications and Networking Conference, WCNC*, 2013. doi: 10.1109/WCNC.2013.6554748.
23. R. Balasubramaniam, S. Nagaraj, M. Sarkar, C. Paolini, and P. Khaitan, "Cell Zooming for Power Efficient Base Station Operation," 2013 9th International Wireless Communications and Mobile Computing Conference (IWCMC), Sardinia, Italy, 2013, pp. 556-560. doi: 10.1109/IWCMC.2013.6583618.
24. J. Lin, Y. Chen, H. Zheng, M. Ding, P. Cheng, and L. Hanzo, "A data-driven base station sleeping strategy based on traffic prediction," *IEEE Transactions on Network Science and Engineering*, 2021. doi: 10.1109/TNSE.2021.3109614.
25. P. González-Brevis *et al.*, "Base station location optimization for minimal energy consumption in wireless networks," in *IEEE Vehicular Technology Conference*, 2011. doi: 10.1109/VETECS.2011.5956204.
26. P. Long, J. Li, N. Liu, Z. Pan, and X. You, "Energy-saving small-cell wake up strategy for ultra-dense networks based on user behavior prediction," in *ACM International Conference Proceeding Series*, 2023. doi: 10.1145/3603781.3603824.
27. G. Chopra, "An efficient base station sleeping configuration for ultra-dense networks," in *2023 International Conference on Emerging Smart Computing and Informatics, ESCI 2023*, 2023. doi: 10.1109/ESCI56872.2023.10100245.
28. F. Han, Z. Safar, W. S. Lin, Y. Chen, and K. J. R. Liu, "Energy-efficient cellular network operation via base station cooperation," in *IEEE International Conference on Communications*, 2012. doi: 10.1109/ICC.2012.6364370.
29. B. Leng, X. Gu, L. Zhang, and L. Zhang, "Energy efficiency optimization in 3D dense urban scenarios with coordinate multi-point transmission and sleep strategy," in *IEEE International Symposium on Personal, Indoor and Mobile Radio Communications, PIMRC*, 2018. doi: 10.1109/PIMRC.2018.8580851.
30. "3GPP Specification: 3GPP TS 32.521," 2010. Accessed: Aug. 17, 2024. [Online]. Available: http:// www.3gpp.org/ftp/Specs/html-info/32521.htm
31. J. Lorincz, "Energy-efficient wireless cellular communications through network resource dynamic adaptation," *International Journal of Business Data Communications and Networking*, vol. 9, no. 2, 2013. doi: 10.4018/jbdcn.2013040102.

32. J. Lorincz and T. Matijevic, "Energy-efficiency analyses of heterogeneous macro and micro base station sites," in *Computers and Electrical Engineering*, 2014. doi: 10.1016/j.compeleceng.2013.10.013.
33. L. Li and W. Meng, "Collaborative base station sleeping solution design in heterogeneous cellular network," in *APCC 2022 - 27th Asia-Pacific Conference on Communications: Creating Innovative Communication Technologies for Post-Pandemic Era*, 2022. doi: 10.1109/APCC55198.2022.9943603.
34. X. Li and C. Zhang, "Semi-dynamic Markov approximation-based base station sleep with user association for heterogeneous networks," *IET Communications*, vol. 17, no. 6, 2023. doi: 10.1049/cmu2.12574.
35. S. Malta, P. Pinto, and M. Fernandez-Veiga, "Using reinforcement learning to reduce energy consumption of ultra-dense networks with 5G use cases requirements," *IEEE Access*, vol. 11, 2023. doi: 10.1109/ACCESS.2023.3236980.
36. N. Piovesan, D. Lopez-Perez, M. Miozzo, and P. Dini, "Joint load control and energy sharing for renewable powered small base stations: A machine learning approach," *IEEE Transactions on Green Communications and Networking*, vol. 5, no. 1, 2021. doi: 10.1109/TGCN.2020.3027063.
37. A. Aggarwal and D. Selvamuthu, "Energy optimization of a base station using Q-learning algorithm," in *17th International Conference on Telecommunications, ConTEL 2023*, 2023. doi: 10.1109/ConTEL58387.2023.10198963.
38. I. Guerra, B. Yin, S. Zhang, and Y. Cheng, "Optimization of base station ON-Off switching with a machine learning approach," in *IEEE International Conference on Communications*, 2021. doi: 10.1109/ICC42927.2021.9500412.
39. P. Kaur and R. Garg, "Improving energy efficiency of 5G base stations: A comprehensive AI-Based optimization approach," in *Lecture Notes in Networks and Systems*, 2023. doi: 10.1007/978-981-99-1699-3_31.
40. T. Pan, X. Wu, and X. Li, "Dynamic multi-sleeping control with diverse quality-of-service requirements in sixth-generation networks using federated learning," *Electronics (Switzerland)*, vol. 13, no. 3, 2024. doi: 10.3390/electronics13030549.
41. T. P. Fowdur and B. Doorgakant, "A review of machine learning techniques for enhanced energy efficient 5G and 6G communications," 2023. doi: 10.1016/j.engappai.2023.106032.
42. J. Lorincz, T. Matijevic, and G. Petrovic, "On interdependence among transmit and consumed power of macro base station technologies," *Computer and Communications*, vol. 50, 2014. doi: 10.1016/j.comcom.2014.02.010.
43. S. Biswas, A. M. Nasir, and M. F. Hossain, "A deep learning based energy efficient downlink power control mechanism for cellular networks," in *Proceedings of 2020 11th International Conference on Electrical and Computer Engineering, ICECE 2020*, 2020. doi: 10.1109/ICECE51571.2020.9393086.
44. Y. Jie, T. Zhenjian, H. Han, and C. Yonghong, "Energy-efficient microcell base station power control in heterogeneous cellular network," in *International Conference on Communication Technology Proceedings, ICCT*, 2017. doi: 10.1109/ICCT.2017.8359697.
45. M. A. Charar and Z. Guennoun, "Energy efficient power control for device to device communication in 5G networks," *International Journal of Electrical and Computer Engineering*, vol. 10, no. 4, 2020. doi: 10.11591/ijece.v10i4.pp4118-4135.
46. R. Kurda, "Heterogeneous networks: Fair power allocation in LTE-A uplink scenarios," *PLoS One*, vol. 16, no. 6 June 2021. doi: 10.1371/journal.pone.0252421.

47. V. Khodamoradi, A. Sali, A. A. Salah, B. M. Ali, R. S. A. R. Abdullah, and I. Krikidis, "Energy efficient base station transmit power adaptation for green 5G massive MIMO systems," in *IEEE Vehicular Technology Conference*, 2019. doi: 10.1109/VTCSpring.2019.8746310.
48. S. Cui, A. J. Goldsmith, and A. Bahai, "Energy-efficiency of MIMO and cooperative MIMO techniques in sensor networks," *IEEE Journal on Selected Areas in Communications*, vol. 22, no. 6, 2004. doi: 10.1109/JSAC.2004.830916.
49. S. Wesemann, J. Du, and H. Viswanathan, "Energy efficient extreme MIMO: Design goals and directions," *IEEE Communications Magazine*, vol. 61, no. 10, 2023. doi: 10.1109/MCOM.004.2200958.
50. J. Zhang, H. Deng, Y. Li, Z. Zhu, G. Liu, and H. Liu, "Energy efficiency optimization of massive MIMO system with uplink multi-cell based on imperfect CSI with power control," *Symmetry (Basel)*, vol. 14, no. 4, 2022. doi: 10.3390/sym14040780.
51. M. Y. Ali, T. Hossain, and M. M. Mowla, "A trade-off between energy and spectral efficiency in massive MIMO 5G system," in *3rd International Conference on Electrical, Computer and Telecommunication Engineering, ICECTE 2019*, 2019. doi: 10.1109/ICECTE48615.2019.9303551.
52. L. Sanguinetti, A. Zappone, and M. Debbah, "Deep learning power allocation in massive MIMO," in *Conference Record - Asilomar Conference on Signals, Systems and Computers*, 2018. doi: 10.1109/ACSSC.2018.8645343.
53. I. Salah, M. M. Mabrook, K. H. Rahouma, and A. I. Hussein, "Energy efficiency optimization in adaptive massive MIMO networks for 5G applications using genetic algorithm," *Optical and Quantum Electronics*, vol. 54, no. 2, 2022. doi: 10.1007/s11082-021-03507-5.
54. V. Kikan and A. Kumar, "A detailed review of 5G MIMO and array antenna design evolution with performance enhancement for mmWave applications," *Wireless Networks*, 2024. doi: 10.1007/s11276-024-03862-7.
55. S. Haykin, "Cognitive radio: Brain-empowered wireless communications," *IEEE Journal on Selected Areas in Communications*, vol. 23, no. 2, 2005. doi: 10.1109/JSAC.2004.839380.
56. O. I. Khalaf, K. A. Ogudo, and M. Singh, "A fuzzy-based optimization technique for the energy and spectrum efficiencies trade-off in cognitive radio-enabled 5G network," *Symmetry (Basel)*, vol. 13, no. 1, 2021. doi: 10.3390/sym13010047.
57. J. H. Huang and S. Y. Hsu, "QoS provisioning in energy-efficient cooperative networks with power assignment and relay deployment planning," *Wireless Networks*, vol. 26, no. 7, 2020. doi: 10.1007/s11276-020-02375-3.
58. I. Al Qerm and B. Shihada, "Enhanced machine learning scheme for energy efficient resource allocation in 5G heterogeneous cloud radio access networks," in *IEEE International Symposium on Personal, Indoor and Mobile Radio Communications, PIMRC*, 2017. doi: 10.1109/PIMRC.2017.8292227.
59. S. Sasikumar and J. Jayakumari, "A novel method for the optimization of spectral -energy efficiency tradeoff in 5 G heterogeneous cognitive radio network," *Computer Networks*, vol. 180, 2020. doi: 10.1016/j.comnet.2020.107389.
60. M. Li, P. Li, X. Huang, Y. Fang, and S. Glisic, "Energy consumption optimization for multihop cognitive cellular networks," *IEEE Transactions on Mobile Computing*, vol. 14, no. 2, 2015. doi: 10.1109/TMC.2014.2320275.

61. S. S. Abidrabbu and H. Arslan, "Energy-efficient resource allocation for 5G cognitive radio noma using game theory," in *IEEE Wireless Communications and Networking Conference, WCNC*, 2021. doi: 10.1109/WCNC49053.2021.9417578.
62. A. Attar, H. Li, and V. C. M. Leung, "Green last mile: How fiber-connected massively distributed antenna systems can save energy," *IEEE Wireless Communications*, vol. 18, no. 5, 2011. doi: 10.1109/MWC.2011.6056694.
63. G. Y. Li *et al.*, "Energy-efficient wireless communications: Tutorial, survey, and open issues," *IEEE Wireless Communications*, vol. 18, no. 6, 2011. doi: 10.1109/MWC.2011.6108331.
64. R. Dangi, P. Lalwani, G. Choudhary, I. You, and G. Pau, "Study and investigation on 5G technology: A systematic review," 2022. doi: 10.3390/s22010026.
65. K. N. R. S. V. Prasad, E. Hossain, and V. K. Bhargava, "Energy efficiency in massive MIMO-Based 5G networks: Opportunities and challenges," *IEEE Wireless Communications*, vol. 24, no. 3, 2017. doi: 10.1109/MWC.2016.1500374WC.
66. M. U. Muzaffar and R. Sharqi, "A review of spectrum sensing in modern cognitive radio networks," 2024. doi: 10.1007/s11235-023-01079-1.
67. J. Lorincz, I. Ramljak, and D. Begušić, "Analysis of the impact of detection threshold adjustments and noise uncertainty on energy detection performance in MIMO-OFDM cognitive radio systems," *Sensors*, vol. 22, no. 2, 2022. doi: 10.3390/s22020631.
68. J. Lorincz, I. Ramljak, and D. Begušić, "A review of the noise uncertainty impact on energy detection with different OFDM system designs," 2019. doi: 10.1016/j.comcom.2019.09.013.
69. J. Lorincz, I. Ramljak, and D. Begušić, "A survey on the energy detection of ofdm signals with dynamic threshold adaptation: Open issues and future challenges," 2021. doi: 10.3390/s21093080.
70. J. Lorincz, I. Ramljak, and D. Begušić, "Performance analyses of energy detection based on square-law combining in MIMO-OFDM cognitive radio networks," *Sensors*, vol. 21, no. 22, 2021. doi: 10.3390/s21227678.
71. I. Ramljak, D. Begušić, and J. Lorincz, "Operational characteristics of square-law combining energy detector in MIMO-OFDM cognitive radio systems," *Applied Sciences (Switzerland)*, vol. 12, no. 9, 2022. doi: 10.3390/app12094684.
72. J. Lorincz, I. Ramljak, and D. Begusic, "Algorithm for evaluating energy detection spectrum sensing performance of cognitive radio MIMO-OFDM systems," *Sensors*, vol. 21, no. 20, 2021. doi: 10.3390/s21206881.
73. M. M. Butt, B. Schubert, M. Kurras, K. Borner, T. Haustein, and L. Thiele, "On the energy-bandwidth trade-off in green wireless networks: System level results," in *2012 1st IEEE International Conference on Communications in China Workshops, ICCC 2012*, 2012. doi: 10.1109/ICCCW.2012.6316482.
74. J. Lorincz, Z. Klarin, and D. Begusic, "Advances in improving energy efficiency of fiber–wireless access networks: A comprehensive overview," 2023. doi: 10.3390/s23042239.
75. S. Shirvani Moghaddam, "Introductory chapter: Primary and secondary users in cognitive radio-based wireless communication systems," in *Cognitive Radio in 4G/5G Wireless Communication Systems*, 2018. doi: 10.5772/intechopen.82033.
76. J. Lorincz, E. Mujaric, and D. Begusic, "Energy consumption analysis of real metro-optical network," in *2015 38th International Convention on Information and Communication Technology, Electronics and Microelectronics, MIPRO 2015 - Proceedings*, 2015. doi: 10.1109/MIPRO.2015.7160333.

77. J. Lorincz and Z. Klarin, "A comprehensive analysis of the impact of an increase in user devices on the long-term energy efficiency of 5G networks," *Smart Cities*, vol. 7, no. 6, pp. 3616–3657, 2024. doi: 10.3390/smartcities7060140.
78. J. Hoydis, M. Kobayashi, and M. Debbah, "Green small-cell networks," *IEEE Vehicular Technology Magazine*, vol. 6, no. 1, 2011. doi: 10.1109/MVT.2010.939904.
79. N. Lemieux and M. Zhao, "Small cells, big impact: Designing power solutions for 5G applications," Dallas, 2019. Accessed: Nov. 13, 2024. [Online]. Available: https://www.ti.com/lit/wp/slyy166/slyy166.pdf
80. A. Jothimani and A. S. Edward, "Review of energy efficient architectures based on small cell driven 5G networks," in *AIP Conference Proceedings*, 2020. doi: 10.1063/5.0028714.
81. H. Jiang, Z. Xiao, Z. Li, J. Xu, F. Zeng, and D. Wang, "An energy-efficient framework for internet of things underlaying heterogeneous small cell networks," *IEEE Transactions on Mobile Computing*, vol. 21, no. 1, 2022. doi: 10.1109/TMC.2020.3005908.
82. S. Samarakoon, M. Bennis, W. Saad, M. Debbah, and M. Latva-Aho, "Ultra dense small cell networks: Turning density into energy efficiency," *IEEE Journal on Selected Areas in Communications*, vol. 34, no. 5, 2016. doi: 10.1109/JSAC.2016.2545539.
83. X. Ge, J. Yang, H. Gharavi, and Y. Sun, "Energy efficiency challenges of 5G small cell networks," *IEEE Communications Magazine*, vol. 55, no. 5, 2017. doi: 10.1109/MCOM.2017.1600788.
84. O. Fratu, A. Vulpe, R. Craciunescu, and S. Halunga, "Small cells in cellular networks: Challenges of future HetNets," *Wireless Personal Communications*, vol. 78, no. 3, 2014. doi: 10.1007/s11277-014-1906-9.
85. Y. Zhang, Y. Xu, Y. Sun, Q. Wu, and K. Yao, "Energy efficiency of small cell networks: Metrics, methods and market," *IEEE Access*, vol. 5, 2017. doi: 10.1109/ACCESS.2017.2696025.
86. J. Lorincz, I. Bule, and M. Kapov, "Performance analyses of renewable and fuel power supply systems for different base station sites," *Energies (Basel)*, vol. 7, no. 12, 2014. doi: 10.3390/en7127816.
87. J. Lorincz and I. Bule, "Renewable energy sources for power supply of base station sites," *International Journal of Business Data Communications and Networking*, vol. 9, no. 3, 2013. doi: 10.4018/jbdcn.2013070104.
88. M. H. Alsharif, R. Kannadasan, A. Jahid, M. A. Albreem, J. Nebhen, and B. J. Choi, "Long-term techno-economic analysis of sustainable and zero grid cellular base station," *IEEE Access*, vol. 9, 2021. doi: 10.1109/ACCESS.2021.3071250.
89. M. L. Ku, W. Li, Y. Chen, and K. J. Ray Liu, "Advances in energy harvesting communications: Past, present, and future challenges," *IEEE Communications Surveys and Tutorials*, vol. 18, no. 2, 2016. doi: 10.1109/COMST.2015.2497324.
90. S. Ozyurt, A. F. Coskun, S. Buyukcorak, G. Karabulut Kurt, and O. Kucur, "A survey on multiuser SWIPT communications for 5G+," *IEEE Access*, vol. 10, 2022. doi: 10.1109/ACCESS.2022.3212774.
91. X. Huang *et al.*, "Software defined energy harvesting networking for 5G green communications," *IEEE Wireless Communications*, vol. 24, no. 4, 2017. doi: 10.1109/MWC.2017.1600360.
92. Y. Liu, Y. Zhang, R. Yu, and S. Xie, "Integrated energy and spectrum harvesting for 5G wireless communications," *IEEE Network*, vol. 29, no. 3, 2015. doi: 10.1109/MNET.2015.7113229.

93. H. Heidari, O. Onireti, R. Das, and M. Imran, "Energy harvesting and power management for IoT devices in the 5G Era," *IEEE Communications Magazine*, vol. 59, no. 9, 2021. doi: 10.1109/MCOM.101.2100487.
94. T. Sanislav, G. D. Mois, S. Zeadally, and S. C. Folea, "Energy harvesting techniques for Internet of Things (IoT)," *IEEE Access*, vol. 9, 2021. doi: 10.1109/ACCESS.2021.3064066.
95. J. Yang, Z. Yang, and Y. Duan, "Optimal capacity and operation strategy of a solar-wind hybrid renewable energy system," *Energy Convers Manag*, vol. 244, 2021. doi: 10.1016/j.enconman.2021.114519.
96. V. Chamola and B. Sikdar, "Solar powered cellular base stations: Current scenario, issues and proposed solutions," *IEEE Communications Magazine*, vol. 54, no. 5, 2016. doi: 10.1109/MCOM.2016.7470944.
97. I. F. Bitterlin, "Modelling a reliable wind/PV/storage power system for remote radio base station sites without utility power," *J Power Sources*, vol. 162, no. 2 SPEC. ISS., 2006. doi: 10.1016/j.jpowsour.2005.07.011.
98. P. Zhao, W. Xu, S. Zhang, J. Wang, and Y. Dai, "Technical feasibility assessment of a standalone photovoltaic/wind/adiabatic compressed air energy storage based hybrid energy supply system for rural mobile base station," *Energy Convers Manag*, vol. 206, 2020. doi: 10.1016/j.enconman.2020.112486.
99. O. Abuajwa and S. Mitani, "Dynamic resource allocation for energy-efficient downlink NOMA systems in 5G networks," *Heliyon*, vol. 10, no. 9, May 2024. doi: 10.1016/j.heliyon.2024.e29956.
100. W. U. Khan, X. Li, A. Ihsan, Z. Ali, B. M. Elhalawany, and G. A. S. Sidhu, "Energy efficiency maximization for beyond 5G NOMA-enabled heterogeneous networks," *Peer Peer Netw Appl*, vol. 14, no. 5, 2021. doi: 10.1007/s12083-021-01176-5.
101. H. Zhang, F. Fang, J. Cheng, K. Long, W. Wang, and V. C. M. Leung, "Energy-efficient resource allocation in NOMA heterogeneous networks," *IEEE Wireless Communications*, vol. 25, no. 2, 2018. doi: 10.1109/MWC.2018.1700074.
102. H. Zhang *et al.*, "Energy efficient dynamic resource optimization in NOMA system," *IEEE Transactions on Wireless Communications*, vol. 17, no. 9, 2018. doi: 10.1109/TWC.2018.2844359.
103. J. Tang *et al.*, "Energy efficiency optimization for NOMA with SWIPT," *IEEE Journal on Selected Topics in Signal Processing*, vol. 13, no. 3, 2019. doi: 10.1109/JSTSP.2019.2898114.
104. A. A. Nasir, H. D. Tuan, T. Q. Duong, and M. Debbah, "NOMA throughput and energy efficiency in energy harvesting enabled networks," *IEEE Transactions on Communications*, vol. 67, no. 9, 2019. doi: 10.1109/TCOMM.2019.2919558.
105. H. Huang, Y. Yang, Z. Ding, H. Wang, H. Sari, and F. Adachi, "Deep learning-based sum data rate and energy efficiency optimization for MIMO-NOMA systems," *IEEE Transactions on Wireless Communications*, vol. 19, no. 8, 2020. doi: 10.1109/TWC.2020.2992786.
106. Y. Zhang, X. Wang, and Y. Xu, "Energy-efficient resource allocation in uplink NOMA systems with deep reinforcement learning," in *2019 11th International Conference on Wireless Communications and Signal Processing, WCSP 2019*, 2019. doi: 10.1109/WCSP.2019.8927898.
107. J. Lorincz, A. Kukuruzović, and Z. Blažević, "A comprehensive overview of network slicing for improving the energy efficiency of fifth-generation networks," May 01, 2024, *Multidisciplinary Digital Publishing Institute (MDPI)*. doi: 10.3390/s24103242.

108. O. Akin, U. C. Gulmez, O. Sazak, O. U. Yagmur, and P. Angin, "GreenSlice: An energy-efficient secure network slicing framework," *Journal of Internet Services and Information Security*, vol. 12, no. 1, 2022. doi: 10.22667/JISIS.2022.02.28.057.
109. F. Zhou *et al.*, "Automatic network slicing for IoT in Smart City," *IEEE Wireless Communications*, vol. 27, no. 6, 2020. doi: 10.1109/MWC.001.2000069.
110. Y. Wu, H. N. Dai, H. Wang, Z. Xiong, and S. Guo, "A survey of intelligent network slicing management for industrial IoT: Integrated approaches for smart transportation, smart energy, and smart factory," 2022. doi: 10.1109/COMST.2022.3158270.
111. M. Masoudi, O. T. Demir, J. Zander, and C. Cavdar, "Energy-optimal end-to-end network slicing in cloud-based architecture," *IEEE Open Journal of the Communications Society*, vol. 3, 2022. doi: 10.1109/OJCOMS.2022.3162116.
112. Y. Xiao and M. Krunz, "Dynamic network slicing for scalable fog computing systems with energy harvesting," *IEEE Journal on Selected Areas in Communications*, vol. 36, no. 12, 2018. doi: 10.1109/JSAC.2018.2871292.
113. C. Tipantuña and X. Hesselbach, "Adaptive energy management in 5G network slicing: Requirements, architecture, and strategies," *Energies (Basel)*, vol. 13, no. 15, 2020. doi: 10.3390/en13153984.
114. International Telecommunication Union, "Measuring digital development: Facts and figures 2020," Geneva: ITU, 2020. Retrieved from https://www.itu.int/en/ITU-D/Statistics/Documents/facts/FactsFigures2020.pdf.
115. Federal Communications Commission, "In the matter of kuiper systems LLC application for authority to deploy and operate a Ka-band non-geostationary satellite orbit system (FCC 20-102)," 2020. Accessed: Feb. 27, 2025. [Online]. Available: https://docs.fcc.gov/public/attachments/FCC-20-102A1.pdf
116. OneWeb, "OneWeb secures investment from SoftBank and hughes network systems." Accessed: Feb. 27, 2025. [Online]. Available: https://oneweb.net/resources/oneweb-secures-investment-softbank-and-hughes-network-systems
117. Q. Wu and R. Zhang, "Towards smart and reconfigurable environment: Intelligent reflecting surface aided wireless network," *IEEE Communications Magazine*, vol. 58, no. 1, 2020. doi: 10.1109/MCOM.001.1900107.
118. J. C. Liang *et al.*, "A filtering reconfigurable intelligent surface for interference-free wireless communications," *Nat Commun*, vol. 15, no. 1, p. 3838, 2024. doi: 10.1038/s41467-024-47865-6.
119. E. F. Ibrahim, *Selected Topics on Reconfigurable Intelligent Surfaces*. Luleå University of Technology, Signals and Systems, Luleå, Sweden, 2022.
120. Y. Liu *et al.*, "Reconfigurable intelligent surfaces: Principles and opportunities," 2021. doi: 10.1109/COMST.2021.3077737.
121. I. P. Hong, "Reviews based on the reconfigurable intelligent surface technical issues," 2023. doi: 10.3390/electronics12214489.
122. C. Huang, A. Zappone, G. C. Alexandropoulos, M. Debbah, and C. Yuen, "Reconfigurable intelligent surfaces for energy efficiency in wireless communication," in *IEEE Transactions on Wireless Communications*, 2019. doi: 10.1109/TWC.2019.2922609.
123. Z. Yang *et al.*, "Energy-efficient wireless communications with distributed reconfigurable intelligent surfaces," *IEEE Transactions on Wireless Communications*, vol. 21, no. 1, 2022. doi: 10.1109/TWC.2021.3098632.

124. Z. Li, J. Zhang, J. Zhu, and L. Dai, "RIS energy efficiency optimization with practical power models," in *2023 International Wireless Communications and Mobile Computing, IWCMC 2023*, 2023. doi: 10.1109/IWCMC58020.2023.10183034.
125. Y. Huo, X. Dong, and N. Ferdinand, "Distributed reconfigurable intelligent surfaces for energy-efficient indoor terahertz wireless communications," *IEEE Internet Things J*, vol. 10, no. 3, 2023. doi: 10.1109/JIOT.2022.3214471.
126. A. V Parambath, J. Flordelis, C. Madapatha, F. Rusek, E. Bengtsson, and T. Svensson, "Integrating Reconfigurable Intelligent Surfaces (RISs) into indoor D-MIMO networks for 6G," in *2024 IEEE 100th Vehicular Technology Conference (VTC2024-Fall)*, 2024, pp. 1–6. doi: 10.1109/VTC2024-Fall63153.2024.10757793.
127. H. Zhou *et al.*, "Hierarchical reinforcement learning for RIS-assisted energy-efficient RAN," in *Proceedings - IEEE Global Communications Conference, GLOBECOM*, 2022. doi: 10.1109/GLOBECOM48099.2022.10001554.
128. Y. Lu, M. Hao, and R. Mackenzie, "Reconfigurable intelligent surface based hybrid precoding for THz communications," *Intelligent and Converged Networks*, vol. 3, no. 1, 2022. doi: 10.23919/ICN.2022.0003.
129. C. E. Worka, F. A. Khan, Q. Z. Ahmed, P. Sureephong, and T. Alade, "Reconfigurable Intelligent Surface (RIS)-assisted Non-Terrestrial Network (NTN)-based 6G communications: A contemporary survey," *Sensors*, vol. 24, no. 21, 2024. doi: 10.3390/s24216958.

5

Future Challenges and Opportunities

With the accelerated development of fifth-generation (5G) networks and the constant demand for ever-higher transmission speeds and the transfer of ever-increasing amounts of data, optimizing the energy consumption (EC) of mobile communication networks becomes a key research challenge for ensuring mobile network long-term sustainability. Until their full implementation, it is predicted that 5G networks will improve energy efficiency (EE) by 100 to 1,000 times compared to previous fourth-generation (4G) networks [1]. This requirement is expected to be met through the implementation of a 5G network based on a heterogeneous base station (BS) infrastructure, while meeting the data transmission demands of the ever-growing number of 5G user devices (UDs). Therefore, one of the key research challenges related to improving mobile network EE is solving the impact of the continuous increase in the number of UDs and corresponding traffic volumes on the increase in the EC of 5G and future network generations. Of particular importance in solving this challenge is taking into account the impact of different approaches to the allocation of new 5G and beyond 5G (B5G) BSs and the need for combining versatile radio access network (RAN) resource management strategies in the 5G and B5G networks.

Furthermore, one of the key possibilities for improving mobile network EE in the future is the integration of local data caching in mobile edge networks (MENs). Data caching at the edge of the network, especially in small BSs and UDs, is becoming crucial to reduce network EC and improve network performance [2]. This is because data caching contributes to the shortening of information transmission paths and reduction of data exchange volumes, which consequently has a positive impact on network EE. However, dynamic changes in the amount of information content transfer, user mobility, and the need for intelligent techniques in deciding on the caching schedule represent significant implementation challenges in future mobile networks. The development of advanced algorithms for caching optimization with the aim of reducing mobile network EC without compromising the quality of service (QoS) in MENs will be necessary. A possible approach to caching optimization and improving the EE of mobile networks is the use of artificial intelligence (AI) and intelligent learning techniques. Research presented in [3] highlights the importance of leveraging machine learning (ML) algorithms in MENs to improve network EE through optimized cache placement according to user mobility patterns. These improvements highlight the potential of

DOI: 10.1201/9781003615866-5

AI-driven strategies to support energy-aware operations in next-generation mobile networks.

Additionally, research attention also needs to be directed to the development of new methods dedicated to improving network EE through optimal allocation of network resources in the RAN according to traffic load dynamics. Finding the optimal compromise between the transmit (Tx) and processing power of BSs is crucial for improving BS EE, and spectral efficiency (SE), especially considering the existence of different transmission technologies in mobile networks. The application of software-defined networking (SDN) provides a potential solution to this challenge, since the SDN concept can enable dynamic control and optimization of network resources [4]. However, further research is needed to explore specific SDN potentials and limitations in the context of improving mobile EE, especially for network implementations realized in combination with MEN and network caching.

Due to the expected vast expansion of Internet of Things (IoT) networks, the EC of such networks will increase and one of the underexplored research challenges in assessing the EE of IoT networks is the impact of data transmission delays on performance and EC of such networks. By taking these delays into account, it is possible to achieve more realistic models and more accurate estimates of the overall mobile network EE. In [5], it is investigated how the optimization of IoT network EE can be significantly improved through the balancing between EC and delay in multi-access edge computing (MEC) IoT networks. Through a stochastic optimization framework, the proposed algorithm addresses challenges posed by user mobility, dynamic processing demands, and varying wireless channel conditions. Simulations show that this approach provides better performance in terms of EE and delay management compared to existing methods.

Also, the network slicing (NS) technique represents a significant opportunity for improving the EE of mobile networks. By applying this technique, it is possible to use network resources precisely tailored to the needs of different use cases (network slices), which contributes to improving the network EE. Possible improvements in network EE when applying NS techniques can be achieved through the optimization of network resource allocation based on the actual resource requirements of a certain network slice, putting a part or complete network slice into sleep mode of operation during periods when there are no active UDs or user traffic, the optimization of shared network resources between different parts of the network through virtualization of network slice elements, and the isolation and grouping of network parts (slices) with different energy consumption requirements. Therefore, the application of NS for improving the EE of mobile networks represents a significantly unexplored area in terms of the possibilities of enhancing the EE of mobile networks.

However, despite extensive research on improving the EE of mobile networks, there is no single approach that comprehensively addresses all aspects of the problem related to the continuous increase of mobile network EC. To enhance the EE of future mobile networks, a comprehensive approach that integrates various solutions across all layers of the Open Systems Interconnection (OSI) model must be envisioned and implemented. Also, initiatives like different collaborative projects such as Green Touch [6], Energy Aware Radio and neTworking tecHnologies (EARTH), 5GrEEn, Opera-Net, Green Radio, and Green-T [7] have addressed specific research challenges in the field of improving mobile network EE and raised awareness of the need for reducing mobile networks' EC. However, the continuation of addressing specific research challenges focused on reducing mobile networks' EC through international joint collaborations on research projects is absolutely necessary in the future.

Also, further development of comprehensive solutions that unify different aspects of network operation in improving EE is needed. A study in [8] highlights the importance of integrating new technologies and network resource management strategies to reduce EC. In [9] the challenge and importance of appropriate trade-offs between EE and SE in multi-operator 5G networks have been emphasized. Thus, future research needs to address efficient balancing among mobile network EE and SE and ensure the sustainable operation of mobile networks while addressing the complexities of high-traffic, multi-operator environments. Furthermore, future optimization frameworks for energy-efficient resource allocation in mobile networks need to address different constraints, including operator-specific service level agreement requirements and possible mobile operators' collaboration on base station site (BSS) and equipment sharing that can result in EE improvements and reduction of operational expenditure (OPEX).

As the demand for energy-efficient 5G networks increases, traditional orthogonal access communication technologies such as orthogonal frequency division multiple access (OFDMA) face limitations in maximizing resource utilization and network EE. A promising alternative is non-orthogonal multiple access (NOMA) technology, which allows multiple users to share the same frequency, time, or code resources by assigning different power levels based on UD location and signal quality. Recent advances highlight the potential of using NOMA in downlink scenarios to improve EE in mobile networks [10]. However, effective optimization techniques for resource allocation within a NOMA system that will address challenges such as user and channel assignment, BS Tx power allocation, and determination of power ratio among UDs need to be developed and standardized. By optimizing these parameters, significant improvements in mobile network EE can be expected. These improvements will be reflected in reducing BS EC compared to conventional access methods such as OFDMA and fractional transmit power allocation [10]. Hence, as 5G networks evolve, integrating NOMA technology with other resource allocation strategies represents a viable direction to further improve network efficiency and sustainability.

Additionally, stochastic geometry has emerged as a valuable statistical framework for analyzing and optimizing the performance of heterogeneous wireless networks. The work presented in [11] highlights the potential of stochastic geometry to improve the understanding and management of EE in heterogeneous networks (HetNets), especially when dealing with the challenges of dynamic spatial deployment and user demand variability. Although this approach has provided significant insight into network behavior, further improvements are needed to enhance its application for EE optimization in complex network environments.

The evolution of the cloud radio access network (C-RAN) has led to the emergence of the open radio access network (O-RAN). It is seen as a revolutionary network architecture dedicated to transforming future mobile networks, especially in the context of improving network EE [12]. Unlike its predecessors, O-RAN introduces open interfaces, virtualization, and intelligent network control mechanisms in mobile network resource management, which enable improved flexibility, scalability, and interoperability among mobile network components. O-RAN facilitates the dynamic allocation of network resources and supports the integration of advanced technologies such as AI and ML [13]. The shift from mobile network architectures relying on proprietary hardware to general-purpose virtualized O-RAN platforms can significantly reduce capital expenditure for mobile network operators and, more importantly, lower OPEX by minimizing the EC of network elements. These improvements, coupled with intelligent network resource orchestration, create opportunities for optimizing BS power consumption and improving network EE, which is essential to address the growing energy demands of 5G networks and beyond [12]. However, despite its potential, O-RAN also presents unique challenges related to EC modeling and the implementation of EE optimization techniques. Researching these challenges is essential to ensure the maximum efficiency of O-RAN while simultaneously reducing network EC, which is a key goal in the development of sustainable and energy-efficient mobile networks. Thus, research related to the development and integration of O-RAN concepts in future mobile networks will be an important contributor to the improvement of mobile network EE.

As discussed in this book, enhancing the EE of 5G and future sixth-generation (6G) networks represents a complex set of challenges. Solving these challenges requires novel solutions in various domains, including optimized mobile network resource management, development of advanced radio access techniques, improved hardware efficiency of network elements, the inclusion of renewable energy sources in the power supply of network elements, intelligent information caching strategies at the network edge, and many more. Although approaches such as AI-driven caching in MENs and resource optimization in HetNets have shown significant progress, the integration of these methods into cohesive frameworks remains an important area of research. Emerging techniques such as network slicing and NOMA offer promising ways to improve EE by addressing specific use case requirements and optimizing resource utilization. Additionally, the development of C-RAN and O-RAN architectures can advance the modularity and interoperability of mobile networks, which are crucial for the improvement of network scalability and EE. However, achieving significant EC reductions in mobile networks will require a comprehensive synthesis of these technologies, tailored to the dynamic requirements of next-generation mobile networks.

5.1 Concluding Remarks

Given the continuous demand for wireless services and the exponential increase in the number of connected devices within mobile networks, the energy efficiency of mobile networks has become a critical factor for the long-term sustainability of technological advancements in next-generation mobile networks. This book therefore thoroughly explores the key strategies and technological solutions that contribute to improving EE in mobile networks, with a particular focus on the evolution of mobile networks toward 5G and the upcoming 6G technologies.

This book describes the motivation for conducting research aimed at improving EE in mobile networks. It also outlines the generational evolution of mobile networks, with an emphasis on development trends and standardization of 5G and 6G networks. The book provides an in-depth analysis of EE in mobile networks, covering metrics for network EE, key factors affecting EE in mobile networks, and BSs as the largest consumers of energy, as well as examining the impact of SDN technology on network EE.

In addition, the book provides a detailed analysis of approaches for optimizing EC within mobile networks. Special attention is given to the RAN, identified as the largest energy consumer within mobile networks. The proposed approaches to energy optimization include improvements in hardware components such as radio frequency power amplifiers and baseband processing units within BSs, alongside the development of efficient algorithms and protocols for implementing advanced radio resource management techniques. In this context, advanced methods for enhancing EE, such as dynamic radio resource and transmit power management, sleep mode techniques, the deployment of heterogeneous mobile networks, network slicing, NOMA technology, and the introduction of satellite-based mobile network constellations are explored.

Despite the substantial benefits of 5G technology in terms of network capacity and performance, this book show that challenges remain concerning the long-term sustainability of mobile networks' EC. These challenges primarily reflect increased mobile network energy demand due to the higher density of network equipment, particularly with the rising number of installed BSs. Future 6G networks promise additional EE improvements through the large-scale application of AI and energy harvesting techniques. Consequently, applying AI and machine learning in mobile networks will play a crucial role in optimizing network operations, enabling real-time intelligent resource management in response to traffic variations to optimize energy use without compromising service quality.

This book also highlights the importance of balancing technological progress with sustainable development goals. On a global scale, the telecommunications sector must adapt to increasingly stringent environmental standards and regulations requiring reductions in greenhouse gas emissions and optimization of EC. Future 6G networks, which will rely on advanced AI systems and energy harvesting technologies, represent a step forward in meeting these goals, though they also present new challenges in implementation and management.

Ultimately, this book underscores the need for ongoing development and implementation of innovative algorithms, technologies, and models that will ensure the long-term sustainability of mobile networks in terms of EE optimization. Future research should focus on further refining existing technologies as well as exploring new ones to achieve optimal EE in mobile networks. Optimizing EE must simultaneously meet growing demands for higher data rates and network capacities, while supporting connectivity for an increasing number of mobile users. Balancing technological advancement with the environmental and economic goals that telecommunications operators must meet will be critical for the future of the telecommunications sector, as it faces sustainable development challenges within an increasingly complex global environment.

References

1. X. Ge, J. Yang, H. Gharavi, and Y. Sun, "Energy efficiency challenges of 5G small cell networks," *IEEE Communications Magazine*, vol. 55, no. 5, 2017. doi: 10.1109/MCOM.2017.1600788.
2. L. B. Mohammed, A. Anpalagan, and M. Jaseemuddin, "Energy and latency efficient caching in mobile edge networks: Survey, solutions, and challenges," 2023. doi: 10.1007/s11277-023-10187-9.
3. L. B. Mohammed, "Latency efficient cache placement using learning techniques in mobile edge networks," Doctoral dissertation, Toronto Metropolitan University, Toronto, Canada, 2022.
4. C. Liang, Y. He, F. R. Yu, and N. Zhao, "Energy-efficient resource allocation in software-defined mobile networks with mobile edge computing and caching," in *2017 IEEE Conference on Computer Communications Workshops, INFOCOM WKSHPS 2017*, 2017. doi: 10.1109/INFCOMW.2017.8116363.

5. H. Hu, W. Song, Q. Wang, R. Q. Hu, and H. Zhu, "Energy efficiency and delay tradeoff in an MEC-enabled mobile IoT network," *IEEE Internet of Things Journal*, vol. 9, no. 17, 2022. doi: 10.1109/JIOT.2022.3153847.
6. U. Barth *et al.*, "GreenTouch roadmap: Strategic research areas and project portfolio," 2012. Accessed: Aug. 23, 2024. [Online]. Available: https://www.bell-labs.com/greentouch/uploads/documents/GreenTouch%20Strategic%20Research%20Areas%20and%20Project%20Portfolio.pdf
7. S. Z. Asif, *5G Mobile Communications Concepts and Technologies*. 2018. doi: 10.1201/9780429466342.
8. A. Clemm and C. Westphal, "Challenges and opportunities in green networking," in *Proceedings of the 2022 IEEE International Conference on Network Softwarization: Network Softwarization Coming of Age: New Challenges and Opportunities, NetSoft 2022*, 2022. doi: 10.1109/NetSoft54395.2022.9844020.
9. O. Aydin, E. A. Jorswieck, D. Aziz, and A. Zappone, "Energy-spectral efficiency tradeoffs in 5G multi-operator networks with heterogeneous constraints," *IEEE Transactions on Wireless Communications*, vol. 16, no. 9, 2017. doi: 10.1109/TWC.2017.2716948.
10. O. Abuajwa and S. Mitani, "Dynamic resource allocation for energy-efficient downlink NOMA systems in 5G networks," *Heliyon*, vol. 10, no. 9, May 2024. doi: 10.1016/j.heliyon.2024.e29956.
11. A. Shabbir, S. Rizvi, M. M. Alam, F. Shirazi, and M. M. Su'ud, "Optimizing energy efficiency in heterogeneous networks: An integrated stochastic geometry approach with novel sleep mode strategies and QoS framework," *PLoS One*, vol. 19, no. 2 February 2024. doi: 10.1371/journal.pone.0296392.
12. A. I. Abubakar, O. Onireti, Y. Sambo, L. Zhang, G. K. Ragesh, and M. Ali Imran, "Energy efficiency of open radio access network: A survey," in *IEEE Vehicular Technology Conference*, 2023. doi: 10.1109/VTC2023-Spring57618.2023.10200477.
13. S. K. Singh, R. Singh, and B. Kumbhani, "The evolution of radio access network towards open-RAN: Challenges and opportunities," in *2020 IEEE Wireless Communications and Networking Conference Workshops, WCNCW 2020 Proceedings*, 2020. doi: 10.1109/WCNCW48565.2020.9124820.

Index

For Product Safety Concerns and Information please contact our EU representative GPSR@taylorandfrancis.com Taylor & Francis Verlag GmbH, Kaufingerstraße 24, 80331 München, Germany